DIMENSION SIX

THE STRUCTURE OF CONSCIOUS EXPERIENCE

SAMUEL AVERY

Wetware Media, LLC

Wetware Media, LLC
wetwaremedia.com

Edited by Margaret Stewart

ISBN: 978-1-954566-05-7

Publisher's Cataloging-in-Publication data

Names: Avery, Samuel, author.
Title: Dimension six : the structure of conscious experience /
Samuel Avery.
Description: Boulder, CO: Wetware Media, 2024.
Identifiers: LCCN: 2023946691 | ISBN: 978-1-954566-05-7
(paperback) | 978-1-954566-06-4 (ebook)
Subjects: LCSH: Physics. | Physics--Philosophy. | Quantum
physics. | Consciousness. | BISAC: SCIENCE / Physics /
Quantum Theory | SCIENCE / Physics / Relativity
Classification: LCC QC174.13 .A84 2024 | DDC 530.12--dc23

Books & Audiobooks
by Samuel Avery

The Quantum Screen:
The Enigmas of Modern Physics and a
New Model of Perceptual Consciousness

Dimensions Within:
Physics and the Structure of Consciousness

The Dimensional Structure of Consciousness:
A Physical Basis for Immaterialism

Buddha and the Quantum:
Hearing the Voice of Every Cell

Transcendence of the Western Mind

The Globalist Papers

The Pipeline and the Paradigm: Keystone XL, Tar Sands, and the
Battle to Defuse the Carbon Bomb

Paradigm Shift: Consciousness, Meditation and
Quantum Physics: A Conversation with Samuel Avery (Audio Only)

Soul of the Kingdom:
Biological Process and the Structure of Consciousness

CONTENTS

PREFACE

This book is written for people with some exposure to physical science: anything from a high school course in basic physics to a PhD. More importantly, it is written for people willing to think new thoughts, as it challenges the most fundamental assumptions underlying our understanding of consciousness and physical reality.

Dimensions are so basic to everyday experience that we assume they are structures of an external world that we are in. But with the arrival of relativity theory and quantum mechanics, this assumption has become questionable. Dimensions that seemed firm, rigid, and rectiliner are now seen to warp, dilate, and blend into one another at their extreme limits.

Modern physics is dazzling. What is it about space that will not let an object move faster than light? Why does mass increase and time slow at extremely high velocities? Why is it impossible to know where a particle is if its momentum is known? In everyday experience, why do we feel acceleration but not velocity? These physical phenomena are well known, and their effects easily calculated, but the dazzle remains. Why do they happen at extreme dimensions—the very fast, the very small, the very massive, the very distant? The dazzle revolves around the role of the observer, or, in other words, consciousness. Without an observer, the existence of an object itself becomes problematic. It is time now to look again at what dimensions are and how they relate to consciousness.

What is a dimension? Is it a fundamental structure of a material universe or a fundamental structure of consciousness?

Will you want to read this book?

If you cannot consider a continuity between introspection and physical reality, this book will not make sense. If you are satisfied that modern physics need not reconsider its fundamental assumptions, or that descriptive language without recourse to basic physical principles can explain away relativity theory and quantum mechanics, this book is not for you. But if you are looking for an understanding of the enigmas of modern physics that explains why we think the world is external and material, you will be interested in, and perhaps enlightened by, *Dimension Six.*

INTRODUCTION

Why can't an object travel as fast as it wants to through empty space?

Dimensions are not as rigid and distinct as we once thought. At extremely high velocities, space shortens, time slows, and mass increases; at extremely small magnitudes, space, time, and mass blend into one another. At extreme distances, space expands at an accelerating rate; and near an extremely massive body, spacetime becomes curved and distorted. The "role of the observer" (consciousness) is a component of physical interactions in both relativity and quantum mechanics where dimensional extremes are encountered. It is reasonable, therefore, to reconsider what dimensions are and how they relate to consciousness.

Is a dimension a structure of an external universe or a structure of conscious experience?

Where we see an object in space is where we hear and touch it. This coordination of sensory experience is due, we assume, to the existence of the object outside of experience, "waiting" to be perceived. But we never experience beyond experience. Actual perception is only of *information* in a dimensional context: seeing, hearing, and touching the object *at the same location in space and time.* Anything additional is a metaphysical assumption. If we accept the dimensional coordination of sensory information but do not assume the existence of a material universe external to perception causing the coordination, we will make better sense of what science has been trying to tell us for over a hundred years.

But is there physical evidence of a correspondence between dimensions and realms of perceptual consciousness? The correspondence becomes evident where it breaks down — where space, time, and mass become indistinguishable. At this point — the quantum level — realms of perceptual consciousness also become indistinguishable. Visual perception becomes tactile where light becomes both wave and particle. From the quantum to the macroscopic level, dimensions come into existence as vision emerges from the minute tactile sensation of retinal cells.

Light is visual consciousness.

Space is a dimensional structure of light: one second of time is c meters of space.

Empty spacetime is potential perceptual consciousness in all five sensory realms.

We cannot see an object moving faster than light because it is light.

1

DIMENSION ZERO

A point is infinitely small and has no dimension. It is not near anything.

2

DIMENSION ONE

A line is an infinite set of points stretching in one dimension. A line segment, like a piece of string, is a finite range of points in one dimension.

A line is the path of light through space. If there is space.

A star is a point of light in the night sky. With space, light is a line between the star and your eye; without space, the star touches your eye.

3

DIMENSION TWO

A plane is an infinite set of points in two dimensions. A sheet of paper is a finite range of points in two dimensions.

The second dimension is not in the first, nor is it separate from the first. It is a set of possibilities coordinated at right angles (orthogonal) to the first.

An area is a finite two-dimensional range of points.

Model of Dimension Two: A Photograph

A still photograph is a model of two dimensions: no motion and no depth. Images on the surface of the photograph are two-dimensional ranges of points.

In a picture or photograph, the depth dimension is foreshortened into the length and width dimensions. Three dimensions are collapsed into two. We do not expect, in the case of a photograph, to touch images where we see them. We know the depth dimension is at right angles to length and width, though we do not see or touch it as such.

4

DIMENSION THREE

Space is an infinite number of points in three dimensions at right angles to one another.

An image with length, width, and depth is a range, or a finite pattern of points, in three dimensions.

An image perceived in space is physically real. The image is actual perception and the space surrounding it *potential* perception: a finite value in an infinite context.

Volume is a finite range of points in three space dimensions.

. . .

Picture yourself shooting at a rifle range. You need to get all three space dimensions right to hit a stationary target. The bullet will travel in a straight line—one dimension—and will hit the target from any distance that is not so far as to involve gravitational drop, so you need not worry about the distance (depth) dimension. The dimensions you must worry about are the horizontal and vertical. If you aim properly in these two dimensions, the bullet will travel straight through the third dimension and hit the target.

. . .

Empty space is empty consciousness.

If space were completely full, perceptual consciousness would have no meaning.

Information is an actual within a potential: a dot that could be a dot or a dash; a yes that could be a yes or a no; a letter or a number that could have been any other letter or number. Information does not exist without a *channel*, or potential; it has to be something that could be something else. The position of hands on a clock is meaningful only if they could be in some other position. Empty spaces on the face of the clock—spaces where the hands are not—give meaning to the spaces that are occupied. A clock that has stopped gives no information; its hands cannot be anywhere else, so you do not know what the time is. A green light at an intersection indicates you may safely proceed because it is not red. The Dow Jones average provides information because the numbers could be other numbers. Images on a computer screen give news or weather information because they could be trillions of other images. The screen is the channel; the images you see or the letters you read are the actual within the channel. That is how you know it is raining in Australia without being in Australia.

Dimensions are information channels — infinite channels, or *potentials.*

A physical image is information — an actual within potentials — a finite range of points in three infinite dimensions.

The meaning of an image in space lies in where it could be but is not. The vibrancy of what we see in space — that which makes an image a conscious experience — depends on the possibility of it being elsewhere. (The word *image* usually means a visual pattern, but I use it here to mean a pattern in any of the five sensory, or perceptual, realms.)

. . .

Now, picture yourself sitting in a dark room hearing music from a record player on your left. In front of you, across the room, is a vase of flowers. To the right is a table with a lamp turned off. Your friend is sitting farther to the right, on the opposite side of the table. You smell the flowers, but since the lamp is off, there is no way to know where the flowers are. But you can tell that the music you hear is coming from the left. Your friend says she also smells the flowers, somewhere, and hears the music coming from the left. When she turns on the lamp, you see the flowers across the room. The words you hear her say tell you that she, too, sees the flowers there. The two of you agree that the music is coming from the left; and when you turn your heads in that direction, you agree that you both see the record player on the shelf. But you do not agree with your friend when she says she sees the lamp on the table to the left. You are across the table and see it to your right.

After some discussion, you realize that you and your friend are observing the same lamp at the same location but from separate points of view in a spatial framework larger than you perceive directly. This observational space is constructed from information you hear in words. The space you have constructed with your friend is not particular to the perceptual space you experience directly. Through your senses, you perceive; through language, you observe.

. . .

This is extraordinary. It seems ordinary but is not. You don't see what your friend sees; you don't know that she is seeing anything at all. Your experience of her experience is limited to sounds — the words she speaks — words that *coordinate* with what you do see. You assume, from the information conveyed in her words, that she is seeing what you are seeing — that she "has" perceptual consciousness the way you do; but that is an assumption. What you experience is the coordination of your perception and her words. To create observational space, consciousness jumps from perceptual experience to a higher level — the level of observational experience — available to all observers, not just to you.

The jump is not logical or scientific. There is no reason to begin to see things through your friend's eyes or to begin to live in an objective space with her. The jump is existential — an irrational equivalence of your perceptual self with her observational self — an equivalence between what you actually see and feel in your body with what she says she sees and feels in her body. This is a leap from individual to collective consciousness; you still see and feel directly, but you have set your perceptual experience on a level with observational experience. To this extent, you have transcended self.

When you are not looking in the direction she is looking, she can tell you about things you do not see directly. When you do look where she is looking, you will see the same things she says she sees where she says she sees them. Her observations become your *potential perceptions*.

Observation is potential perception.

To create observation from perception, an observer "factors out" her own position (and motion) in space, creating a new potential for information. If you go where she is but do not see or hear what she says she sees or hears, what she says is not potential perception and therefore not observation. It does not coordinate dimensionally with perception.

You see, hear, and smell images in space, at a distance from your body. Perceptual space moves with your body, your head, or just your eyes. But to reconcile what you perceive with what your friend says she perceives, you construct observational space by factoring out your location and hers. When you turn your head to the left (the direction from which you heard the music) to see the record player on the shelf, you rotate the spatial orientation of everything you see and hear in the room to the right. You rotate perceptual space. But the observational framework you constructed with your friend does not rotate.

In perceptual space, the room moves, in observational space, your head moves.

It is easier to move you through space than to move the rest of the universe through space in the opposite direction. Creating a common space for you and your friend to live in creates an opening in the self to a new level of being. Once opened, you are never again only you. The body leaps into observational space, becoming your body in relation to others. New openings broaden to a wider universe of being and doing. (More of this later.)

Patterns of air molecules touch cells in your ears, patterns of photons touch cells in your eyes, and airborne molecules are chemically "tasted" by cells in your nose.

Everything you perceive is transmitted to you through one type of cell or another. A physical image is a pattern or range of minute sensations from millions of sensory cells. Cells do not perceive at a distance — *you* do. Space is the template that arranges the points into a pattern. A single cell in the retina of your eye is touched by a photon, but the cell does not see the pattern produced by many cells touched by many photons. The picture in your mind is composed of information collected from a multitude of sensory cells against a background of empty space. Your mind creates space dimensions to organize cellular sensations into patterns that only the organic wholeness of *you* can perceive. Perceptual consciousness is a composite of cellular sensation; it is reducible to the individual cellular level but experienced by the organism as a whole on the macroscopic level. Cells touch and taste; only *you* see, hear, and smell. Perceptual consciousness is not contained by a cell or group of cells; you, as a sentient multicellular organism, are greater than the sum of your cells.

Space dimensions are infinite, while the width, length, and depth dimensions of an image in space are finite. The infinite dimensions — not the finite — are potentials for sensory information. Each point in the range of points that constitutes the image is a point of potential perception in all five sensory realms. Therefore, where you *actually* hear an image is where you *potentially* see or smell or touch it. You may not actually see or smell or touch it, but because the potentials for seeing, smelling, and touching are coordinated, you potentially perceive the image in these realms at that location. You will touch it if you reach out your hand. This is what makes images in spacetime seem material.

Only the potentials are coordinated; the sound of an image has no bearing on what it looks or smells like. But the shape of the image — its range of points — indicates the range of potential perception. If you see a long, narrow image like a baseball bat, you can touch it at either end or anywhere in the middle, but you do not know from seeing it what it will feel or smell like, only where it may be felt or smelled.

Sensory potentials are related in terms of their emptiness.

The human senses of taste and smell seem insignificant when compared to the richness of sight, touch, and hearing. Our primary consideration here, however, is of potential rather than of actual perception — of the dimensional structure of perceptual consciousness as a whole rather than of any particular experience within that structure. Actual experience in the realms of taste and smell may not be of great physical significance for people, but their potentials mean that we are conscious in these realms. Information within them must be coordinated with the more important realms of seeing, touching, and hearing. The dimensional structure of human perceptual consciousness reveals a mammalian past.

Information in the realms of seeing, hearing, and smelling comes in a form that individual cells do not themselves experience. Cells "touch" on their outer membranes, and "taste" within their membranes but do not see, hear, or smell in, on, or at a distance from the membrane. The three perceptual realms that require space do not exist on the single-cellular level. Space is a creation of multicellular life.

The three realms of exclusively multicellular perception are experienced in three space dimensions. Each is distinct from the others, but coordinated and interchangeable with the others. Turning to see an image where (and when) you hear it creates an illusion of the image waiting in space when you were not looking at it.

Potential perception creates the illusion of independent material existence.

. . .

In the room, as you discuss with your friend the things you each see, hear, and smell, you find common experience within these spatial realms. Your direct experience is more intense and detailed than her descriptions. Your experience of her experience is limited to what you hear her say. The sound you hear — her language — is distinct from the sound of the washing machine or traffic in the street in that the information it carries coordinates with what you see directly. Her spoken observations come to you through the auditory realm, but they are not just noise — they are information that allows you to see through her eyes and hear through her ears. You both see the "same" images only because observational information is coordinated with perceptual consciousness.

. . .

Model of Dimension Three: A Hologram

A three-dimensional hologram is a model for spatial images without mass. Holographic figures are constructed of light alone. There is no substance to them other than light, and one feels no touch sensation at the range of points they occupy. There is no "thing" out there, in space, "causing" perception.

Unlike real space, the hologram does not coordinate visual with other realms of perceptual consciousness. We feel it strange to not touch a holographic figure where we see it. But the fact that we *expect* to touch it there reveals the underlying coordination in the real world of seeing, touching, hearing, tasting, and smelling. We perceive everything in the real physical universe *where we would see it*, that is, in the dimensional structure of light.

View in Three Dimensions

The perceptual picture you experience is of images at ranges of points against a background of empty space. The images could be anywhere else in space. That is the dynamic of perceptual consciousness.

. . .

As you look south you see a field of grass and a pasture beyond the fencerow. Shifting your view to the west, you see a house and a road. You have rotated one space dimension into another. The entire universe has shifted. Or more simply, you have turned your head.

5

DIMENSION FOUR

Spacetime is an infinite set of points in three dimensions of space and one of time.

Time exists at "right angles" to all three space dimensions.

A point in four dimensions is an *event*.

A physical image in motion is a range of events in three finite space dimensions, length, width, and depth, and one finite time dimension.

A velocity is a finite cross-section of space and time.

. . .

Suppose the target you are trying to shoot with the rifle is a glass bottle swinging on a string. The bullet follows a straight path, as before, so you need not worry about the distance dimension. You need to get the horizontal and vertical dimensions right, as before, but now you must also squeeze the trigger at the right time to hit the bottle. This is how to locate a point, or event, in four-dimensional spacetime.

To meet a business associate downtown, you specify two cross streets, the building floor you will be on, and the time. If he is waiting at the wrong intersection or on the wrong floor or at the wrong time, there is no event.

· · ·

An image is a contiguous range of points in spacetime. It can be understood as a *thing* that moves through external space, or as a pattern that moves across a background of points, like the image on a pixel screen. The degree to which it maintains its identity (remains "the same" image) is the degree to which its finite dimensions remain contiguous through space and through time.

· · ·

The coordination of time and space dimensions is demonstrated by moving your body. If you turn your head, you interchange one space dimension for another. If you walk across to the vase on the other side of the room, you interchange time with one space dimension. The perceptual universe moves in the opposite direction at a rate determined by the time it takes you to get to where you are going. The observational space you constructed with your friend does not move when you do.

*With space and **time**, you can walk to the vase and touch it where you see it. The flowers smell stronger up close. You can move to and touch everything you see in the room exactly where you see it. If an image is in motion, you touch it where and **when** you see it. The image becomes not only a range of points in space but a range of events in spacetime. If you touch something without seeing it, you can turn to see it where and when you touched it. Your friend has the same experience you have as she walks about the room. or at least you hear her say that she does. Her words coordinate with what you see and touch and there is no reason to doubt what she says.*

With a few simple scientific instruments that extend perceptual consciousness, you can locate the vase of flowers in space even with the lights off by interchanging, or "rotating," the axis of time into space. As you move about the room using your instruments to measure the magnitude of the flowers' scent at separate events in space and time, you establish a gradient that increases as you come closer to the vase. Walking up the gradient in the dark you will, in time, be able to touch the flowers without ever seeing them. In this manner you construct space from time.

. . .

Space is constructed from time.

Constructing space from time is easier to do with the record player than with the flowers. Sound, unlike smell, reveals the direction of its source — the dimensional structure of sound waves builds a gradient into auditory information, so you know from the sound waves themselves which way to walk, even in the dark. To locate the source of a sound, you walk up the gradient, hearing the sound increase in volume as you go. The space dimension you have used for direction to the source is also constructed from time in that there are slight time differences between the arrival of wave crests in each ear.

Space is constructed from time: we have this from a reliable source. For over a hundred years we have known that space and time are of the same primal substance. On the most fundamental level, they are not separate. c is said to be the "speed" of light (about 300 million meters per second), but it can also be an expression of the identity of space and time. c means one second of time *is the same thing* as 300 million meters of space. Space and time are interwoven into *spacetime*. For example, consider the sky on a starry night. The image you see of a star 100 light-years distant is the image of that star as it was 100 years ago. You are looking back in time: 100 light-years of distance are the same as 100 years of time. Even when you look at the vase across the room, you are looking back in time.

Space is constructed from time, in the form of light.

The same reliable source noted that c remains constant for all observers, *regardless of their velocity relative to each other*. This shows that *observational* spacetime is constructed from *perceptual* spacetime. (And by extension, that "objective" physical dimensions are derived from actual perceptual experience—vision.) Across the table, when there is little or no relative velocity, observational spacetime is constructed by factoring out the location of each observer in space; but at near-light relative velocities, location must be factored out in space*time* (through the Lorentz transformation).

. . .

To your relief you find that you and your friend on the other side of the table always agree on what time it is. You will remember that you did not initially agree on where the table was in space: you saw it to the right, and she claimed to see it to the left. Only after some discussion did you realize that the record player was farther from her than from you, and that the vase, though the same distance from both of you, was at a slightly different angle from each. The problem for observational space, you both came to realize, was the distance between your bodies. But time seems to have a built-in observational quality, at least in the room. Even as you move about, time seems to pass at the same rate everywhere. As long as relative motion remains below several million miles per hour, watches continue to tick at the same rate. So, for all intents and purposes in everyday life, time is just time. Only at a dimensional extreme—only if your velocity relative to one another were to approach the "speed" of light—would time pass at different rates for each of you. In other words, the sameness of time on the everyday level in the room begins to unravel only when the meters in every second of your relative velocity approach the number of meters in every second of light. At such an extremely high velocity you would strain the structure of light and the dimensional structure of spacetime itself.

. . .

At high velocity the difference in time exists only for observational time—only in the time she says she experiences but not in the time you experience directly. In observational spacetime, *c* is a relation (*c* meters per second); while in perceptual spacetime, *c* is an equation (one second = *c* meters).

. . .

For safety's sake, we should not attempt to illustrate this in the room where you are sitting. Flying around at several hundred million meters per second is unsafe indoors, so let us move, temporarily, to a thought experiment in interstellar space. There are no stars, galaxies, or anything else for light-years in any direction. You are still next to your friend, at about the same degree of separation, but this time she is passing by you in her chair at 150,000,000 meters per second, or one half the speed of light. For a split second—a very split second—you are at each other's side as she passes, and due to the wonders of the thought experiment, you are able to compare notes. You have to talk fast.

First, you compare velocities. Your speedometer says 0, which makes sense, as there are no images around you (except for her) and you do not have the sensation that you are moving. When you point your radar gun at her, it measures her velocity at ½ c. But she says her speedometer reads 0, and her radar gun when pointed at you reads ½ c. This seems strange at first, but after some discussion, you realize that there is no such thing as absolute velocity; the velocity each of you measures is relative. As before in the room, taking relative motion into account, you are able to construct an observational framework of spacetime that works for both of you. So far, so good.

Then you look at your clock and at hers. Your clock is ticking at the normal rate, but hers is ticking more slowly. She, however, does not agree with what you see. She says your clock is ticking slowly and hers is perfectly normal. This is disturbing: time seems to be relative, too! The passage of time depends on whose reference frame you are looking into and the degree of relative velocity between your reference frame and theirs. Relative space is fairly easy to conceptualize; you simply abandon the idea that any one person sees the framework for the whole universe. Relative time, though? How can time move more *quickly at one velocity than at another? And that is not all. Space also has problems that you did not notice before.*

You each extend a measuring stick in the direction of your relative motion. Yours is a full meter long, but hers is shorter! Each of her centimeters is shorter than yours. And she mentions, as she swishes by at several hundred million miles per hour, that it is your meter stick that is shorter, not hers. The space dimension in the direction of motion has contracted, as well as time, but only for the other observer. More disturbing yet, her **mass***, and the mass of every object traveling with her, is more than the mass of the identical objects at rest around you! She continues to maintain that the opposite is true. You are heavier, she claims, than you ever were when sitting still in the room. (Not bigger, just heavier.)*

· · ·

What is going on here? In the room, it was easy to reconcile her perceptual experience with yours. All you had to do was factor out the space between her seat and yours. It is more difficult, you find, to construct observational spacetime at half the speed of light. But you can do it. You are both observers — if you take the dilations you have experienced in time, space, and mass into account, you can find a way to construct, or re-construct, an observational spacetime that you both agree on.

. . .

  After your friend passes by a few more times and you continue to compare notes, you realize that the dilations are **all the same**. *This is important. Dilations in space, time, and mass (shorter, slower, more massive) are all by exactly the same factor. For a given relative velocity, space shrinks by the same amount that time slows and mass increases. For slower velocities, as in walking about the room, the dilations are so small as to be entirely unnoticeable — even unmeasurable — while at higher velocities the dilations become extreme. At the speed of light itself, the measuring stick shrinks to zero, time comes to a complete stop, and mass becomes infinite. Nothing can actually move at the speed of light because an infinite amount of energy would be required to accelerate an infinitely massive object the last meter per second to c. No object with any mass at all can ever travel at that velocity. An object can* **approach** *the speed of light, growing shorter, slower, and more massive, but never achieve it. For some reason, a physical object cannot travel as fast as it wants to through empty space. The "speed" of light will not let it.*

. . .

  But what does light have to do with the structure of spacetime?

It is important to remember that we are not talking about optical illusions here. The meter stick does not just *appear* to shrink—it really does shrink. Time does not just seem to pass more slowly—it really does. If *c* were just the "velocity of light," these dilations would be light playing tricks on us. But because *c* is the structure of *spacetime itself*, it is spacetime that shrinks and slows, and mass that increases. Spacetime is derived from light—from actual visual consciousness—but has become the dimensional structure of perceptual consciousness as a whole. Our reliable source did not say as much, but he saw it.

Light is visual consciousness.
The structure of light, *c*, is the structure of spacetime.

Here's how you construct a common framework of space, time, and mass for two observers moving at a relative velocity. (If you don't like mathematical notations, just skip through the following. But the math here isn't terribly hard, so read on if you're feeling adventurous.)

The formula below (the Lorentz transformation) describes the difference between what observers see when they are traveling at relative velocities. Space contracts, time slows, and mass increases by the following factor:

$$\frac{1}{1-\dfrac{v^2}{c^2}}$$

where v is the relative velocity between observers and c is the speed of light.

Let's say the relative velocity is one half the speed of light:

$$v \text{ is } \tfrac{1}{2}\,c, \quad \text{so...} \; v^2 \text{ is } \tfrac{1}{2}\,c \times \tfrac{1}{2}\,c = \tfrac{1}{4}\,c^2$$

$$\text{So... } \frac{v^2}{c^2} \text{ is } \tfrac{1}{4}$$

$1 - \tfrac{1}{4} = \tfrac{3}{4}$, and the square root of $\tfrac{3}{4}$ is .866
$$1/.866 = 1.15$$

At $\tfrac{1}{2}\,c$, you and your friend will each see everything in the other's reference frame as 1.15 times shorter, slower, and more massive than in your own. Her meter stick appears to you only 87 instead of 100 centimeters long. A noticeable difference.

If we do the same calculation at a much slower v, say 1% of c, (still pretty fast: about 6.7 million miles per hour) $1 - v^2/c^2$ becomes .9999 and the dilations shrink to a factor of 1.0001. The meter stick is only a tenth of a millimeter shorter. The dilation, though measurable, is not noticeable, even at millions of miles per hour.

But if v is .99 c, the factor becomes 7.08. Your friend's meter stick will be only 14 centimeters long, her minute will last for 7.08 minutes, and she will weigh close to a thousand pounds. Noticeable.

Whether or not you chose to follow the math, here's the takeaway: at high velocities, the space, time (and mass) dimensions of your perceptual consciousness no longer agree with those of another observer. Observational consciousness cannot be constructed by simply factoring out the location of observers in space; you must factor out their location in space*time* as well, that is, their velocity. Observation, though it coordinates with perception, is not the same as perception. (There are no dilations in the perceptual consciousness of your own reference frame.) For this dimensional extreme, you construct a common universe with other observers (observational spacetime) only by multiplying dimensions by the Lorentz transformation discussed above.

Even under normal conditions, we experience perception in a manner totally distinct from observation. Perception is information seen, touched, smelled, tasted, or heard directly. Observation is *information from observers*. You are aware of observational consciousness through what you hear your friend say she sees (or through pictures she shows you).

What she says she sees, you may not see yourself. If it is truly observational, you would see it too, if you were in her reference frame — if you were sitting with her in her reference frame as she zooms by. What she says is observationally true only if any observer would see the same thing — only if what she says she sees is *potential perceptual consciousness*. If she is hallucinating and says she sees a kangaroo on the ceiling, no one else will say they see it there, and her information is not potential perception. If she is lying, no one else will see what she says she sees. So why do we believe her if she says the dilations are in your reference frame and not in hers? We believe her because a consensus of observers — not just this one friend — will, under the same conditions, report they see the same thing. That's how science works.

Through repeatable experiments we know what she says is right. In the real world we have seen that extremely fast aircraft with extremely accurate clocks measure the time dilation we have calculated. And we have seen that radioactive particles traveling near the speed of light decay more slowly than they would at rest. They also gain mass according to the transformation factor. We have not yet accelerated a living observer to half the speed of light, but we do know what she would see. Enough experiments have been performed and enough observers have agreed on what they see to assure the scientific world that the dilations exist physically.

Science is the systematic construction of observational from perceptual consciousness.

• • •

Back in the room, you and your friend were able to locate images before she turned the lamp on. Without light you could tell there were flowers somewhere in the room and music coming from the left. Space dimensions were in effect. When the light came on you could see exactly where the vase and the record player were, and you realized that where you smelled and heard them beforehand was where you **would have seen** *them with the light on. Light—visual consciousness—has some sort of physical predominance over sound and smell. It may be a matter of acuity—perhaps we see more clearly than we hear or smell—and for that reason give light a preferred position in our overall sense of perceptual consciousness. Or the predominance of light may be due to its capacity to reveal more information. If this were the case, the medium for light waves would exist along with the medium for sound waves and other sensory information but would not constitute spacetime itself as an underlying framework for the whole of perceptual consciousness. Light would have no special status.*

• • •

The predominance of light is more than quantitative. Scientists for over a hundred years have been looking for a medium for light—some sort of material plenum like water or air that carries light waves through space—but they have never found one. Light is not a wave *of* anything. It does not exist *in* anything material. There is no underlying substance between a visual image and an observer. Air molecules, we know, are a vehicle for olfactory particles wafting across the room, and waves of air molecules the substance of sound; but there is no known substance for light. Olfactory and auditory information are reducible to motion in air, but light, it seems, is not reducible to anything. There is nothing more fundamental (macroscopically). Dimensions associated with sound and olfactory sensation are incorporated within the dimensional structure of light, but light itself is not located in anything — not even in space. *Space*, rather, *is in light*. Light determines the dimensional structure not only of itself, but of all the universe.

Which is interesting. One usually assumes that light exists, along with all physical things, in a rigid, external spacetime universe. But the dilations show that whatever spacetime is it is not rigid. It shrinks, slows, and warps under extreme conditions, and the fact that spacetime takes on the structure of light means there is nothing outside of light that light is in. If there were such a structure, we would never be able to see it, or hear it, or touch it. To hold together our concept of the universe, we might believe that some sort of medium for light *has* to exist "out there," but there has never been experimental evidence for it. Scientists looked repeatedly for a medium for light (the ether) in the late 1800s, but never found it.

What we see is a function of how the retina creates millions of cellular sensations that are assembled into a picture of physical reality. This is illustrated by the visual experience — or what we know of the visual experience — of a small crustacean called the *mantis shrimp* (Yong, Ed, *An Immense World*, 2022). Most animals have two classes of cones, or photoreceptors, in their retinas. Humans have three, which gives us a more colorful universe. Many birds have four, which gives them an even wider range of colors and a universe of light and color we humans cannot imagine. (What would another color look like?) The mantis shrimp has twelve kinds of photoreceptors, and lives in a universe entirely distinct from our own. (What would nine more colors look like?) But the mantis shrimp's brain is small, and it seems unlikely he would be able to resolve images in a spatial context the way we do. Rather than seeing separate shapes that he can identify as food, predator, or mate, he may see only separate colors. Where we would see an object in a dimensional relation to other objects, a mantis shrimp may only see patches of color that correspond only distantly to what we would call color. (Another part of its retina may resolve vague images in black and white.) We have no way to know, of course, what the mantis shrimp actually sees, but it is safe to say that his "universe" differs from our own.

Light always moves at c. It never exists at rest and does not accelerate from 0 to c. Light attains the speed c instantaneously, without passing through any intermediate speed. It has no mass (though, strangely, it does have *momentum*, which, macroscopically, is mass times velocity), and light is not subject to the passage of time. From its own reference frame, light is always now. We define a line in space as the path of a ray of light, then turn around and try to find light *in* space! Looking for light in space is like looking for a house in a room of the house we are looking for.

Understanding time as a dimension expands our understanding of what a dimension is. In the room with your friend, under normal conditions, space and time are clearly separate: 300,000,000 meters is such a long distance that we do not confuse it with a second of time. At extremely high velocities, however, space and time become less distinct, and their fundamental identity is revealed in the spacetime relation that is velocity: meters per second. The fact that space is in light, and not light in space, reveals a fundamental relation between dimensions and perceptual consciousness. If light is visual consciousness, dimensions are within consciousness and not consciousness within dimensions.

For observational consciousness, light is a velocity: c meters per second.

For perceptual consciousness, light is an equation: one second equals c meters.

Model of Dimension Four: The Photon Screen

If a subject is shown a red spot on a screen for a split second, followed by a nearby green spot 200 milliseconds later, he invariably reports not two distinct nearby spots, but a single spot *moving* from one location to the other, changing color somewhere in between. (Koler's color phi phenomenon.) Two motionless points are shown in succession, but the subject sees a single *thing* in motion. If the distance or the time interval between spots is increased, the subject begins to see two separate spots of different colors. The single image of one *thing* vanishes, through distance, into two separate *things*. In a similar manner, if you look very closely at a picture on a pixel screen or newspaper photograph, the image disappears into the space between the dots of which the screen or photograph consists. The forest disappears between the trees.

A model for perceptual consciousness in four dimensions is the *Photon Screen*: a three-dimensional screen where images on the screen move. This, of course, is vision itself, but in what sense is it a "screen?" Visual consciousness is like a screen in that it consists of millions of tiny dots of light that images move across. Images appear smooth and continuous on the macroscopic level but coarse and discontinuous on the quantum level—the dots begin to show. Space is "grainy"—if you look too closely (extremely closely) images do not move smoothly through space. As an image moves, it "leaps" from one set of dots to another, like an image on a pixel screen. A high-definition pixel screen has so many points so close together that images seem to move smoothly; you must slow them down, or look very closely, to see the graininess of the pixels. The Photon Screen is millions of times smoother than a pixel screen, which is why nobody noticed the graininess until the early twentieth century.

To envision the Photon Screen, you may experience it within, as a stimulation of cells in the retina. This cannot be done in a scientific manner. Science loses its objectivity straying into conscious experience. We will go beyond the limits of science, therefore, (briefly) to gain a fuller view of what science is, from the outside. We will look within "subjective" experience directly at the process by which light becomes visual consciousness. Could it be that problems in modern physics originate in trying to see light in space, and in failing to appreciate the composite nature of visual consciousness?

Here is a simple way to see the fundamental structure of light directly:

Close your eyes and look at your retina — at your field of vision, without the vision. The shape of the field is roughly that of a horizontal oval, fading into darkness toward the edges. Beyond the edges is nothingness, a void that is not even empty space. Some light will find its way through your eyelids and into the oval, but look past it if you can. Blotches of light and shade may linger as negative images of objects you were just looking at, but this is not what we are interested in. Look past the streaks and blotches to the tiny dots beyond them: thousands of tiny points flashing on and off across the length and width and depth of the oval. This is your *field of vision*—or the Photon Screen—without the photons. The dots are signals from retinal cells as they are arranged into vision. Your visual experience, as a multicellular organism, is a composite of signals from these cells.

Multicellular experience is a composite of cellular experience. That is why spacetime is grainy, or *quantized*.

The Screen is not real in the physical sense. You cannot reach out and touch the little dots. You do not see them the way you see images in the room around you, and some people don't see them at all. (The retina can barely detect a single photon; neural fibers often require a half-dozen or so photons within 100 milliseconds before triggering a signal to the brain.) I cannot say, therefore, that the dots are *there*. You may see them only because I have suggested they exist. The problem, I believe, is that the little flashing dots you may or may not see in your field of vision are the transition between the *subjective* and the *objective*. Neither of us has access to another's subjective experience of the Screen, but we do have access to the images another *says he sees* on the Screen.

Watch the dots for as long as your mind is able to concentrate. They are spread out across the oval of your field of vision and are most clearly experienced in the middle. There is no space between them, and as they flash on and off, there is no time between on and off. They are so small and indistinguishable from one another that it is impossible to focus on any one of them. You are not, after all, seeing them with your eyes.

A signal from a retinal cell is the particle nature of light.

Now, slowly open your eyelids less than halfway. At the top of the oval, you will still see the dots—the Screen itself without images—while at the bottom you will see photons of actual light arranged in dimensional images. Here, at the lower portion of your field of vision, the Screen is no longer visible behind images on the Screen. You are looking at the transition between cellular and multicellular consciousness. At the top of the oval, each dot is the experience of a single retinal cell. At the bottom of the oval, you see physical images—photons arranged in spacetime. The transition between top and bottom—between cellular and multicellular—is the transition between quantum and macroscopic reality, between the particle nature of light and its wave nature, between subjective and objective experience, and between the mind and the physical world.

Photons are dots of light. The mind uses the Screen to assemble cellular experience into a dimensional picture. (Photons fly around in space—or appear to—only in observational consciousness, where there is no perceptual point of view.) You may think of the field of vision—the Photon Screen—as in *space* but try, if you can, to think of it as not in anything. Look at it as it is. Extend confidence, if you are able, to immediate perception without trying to see anything that is not there. Let visual consciousness exist on its own, without context.

The Photon Screen is like a three-dimensional movie screen. It includes time and space dimensions, like a movie, but unlike a movie, the Photon Screen coordinates the visual realm with all perceptual realms (not just the soundtrack). The photons are stacked in a dimensional order that coordinates visual consciousness with the auditory, olfactory, chemical, and tactile consciousness. By interchanging time with space, you "scroll" through the Screen, appearing to move through spacetime.

· · ·

You stand up from your chair and walk out the front door. As you walk, everything you see in space moves in the opposite direction. On the sidewalk in front of your house you see a table in the center of your field of vision. Your friend, who left the house earlier, is approaching from the right on a unicycle, holding a bell in one hand and a chocolate cake in the other. She rings the bell once. As she rides by, she places the cake, somewhat awkwardly, on the table in front of your house and rings the bell again. Before disappearing to the left, she rings the bell a third time. You smell the cake.

You hear the bell and smell the cake from a distance; you do not have to move to perceive them. But to touch the cake or to taste it, you have to move your body. You must rotate time into space. Touch and taste, unlike seeing, hearing, and smelling, are not "spatial."

*As you look and move around the room and walk out the door, you notice that space moves through the horizontal oval of visual consciousness — up-down, right-left, and back-forth — as you interchange each space dimension with time. All images move when you step forward or turn your head. It would be easier to understand that it is you that moves, or just your head, and not the house and the street and the rest of the town: easier to think of moving one thing — your body — than of moving everything else in the universe. But there is no physical reason, at this point, to **not** move everything else in the universe. Images in four dimensions have no weight, and do not resist movement in any way. They are not **objects**.*

*In the room, looking through the oval field of vision, you see dimensional relations within and among images; the vase is at the far side of the room: flower stems are still in the vase with blooms at the end of each stem. To the left, you see the shelf with the record player above. But when you look where the record player is you are no longer looking at the center of the room and do not see the vase and the flowers. To connect what you see in the moment with what you saw before, the mind extends spatial relations in the shelf and record player to those seen a minute earlier in the vase and the flowers. In doing so, you extend **potential** perception beyond actual perception. Through time, you dimensionally connect previous experience to present experience. What you extend beyond the oval is potential perception, because when you scroll back to* *the center of the room you know you will see the vase and the flowers right where you saw them before. Potential becomes actual. Space originates in actual perception—in the light in the oval of the Photon Screen—but becomes larger than actual perception. You don't actually see the vase when you are looking at the record player. The space that extended to the larger room is an extension of the space in the oval. Potential vision is dimensional extrapolation of actual vision.*

. . .

Images reappearing in the oval at a later time create the sense of a rigid external structure of space*time*, though you have no direct experience of such an external structure. You never see it or touch it; all you actually experience is the sensation of seeing and touching. Actual experience is of an oval of light bouncing around the room from vases to tables and record players and shifting from unicycles to tables and chocolate cakes. The mind connects the dots on the Photon Screen through time by constructing a larger dimensional picture of potential seeing, hearing, tasting, smelling, and touching. The extension of the dimensions of the Photon Screen beyond the Screen itself becomes infinite spacetime.

And it is not just your own experience. You are alone in your perceptual sphere, but not in your *potential* perceptual sphere.

. . .

When your neighbor in the house to the left steps out of his front door and the two of you chat, you find that information in his words coordinates with your direct perceptual experience in the oval. He says he saw your friend on the unicycle, saw and heard her ring the bell at three separate events in spacetime, and saw and smelled the cake on the table in front of your house. It seems logical to assume that he, too, "has" an oval. His words, symbols, gestures, and numbers are such good observational information that you assume he is perceiving the same things you are. You have no access to his oval, you cannot confirm that he even has one, but it seems he has to have one in order for you to construct an observational universe with him.

But does he? All you experience is hearing the sound of his words.

. . .

The spacetime universe you construct with your neighbor is of the same dimensional structure as the Screen you extend from the oval of your field of vision, but the quality of the information is not the same. You do not experience the universe he experiences as images on a screen; you experience it as words and symbols. The information from him is a far larger universe than the oval, but not as detailed.

When you turn your head from side to side, or walk out the front door, you see what you saw a little earlier, and if you look, you see what he says he is seeing now. Dimensions are extended down the street and across town, over the oceans and beyond the planets and galaxies — a universe is constructed where actual experience may or may not be located. True, the dimensions are mostly empty, but if a reliable observer were to say he sees or touches anything anywhere or anywhen within them, you would see it there too, if you were to go and look. And you know, because he is a good observer, that what he reports is true, even if you don't look. Because the observational universe you construct with your neighbor (and with everyone else in the world) is so much larger than the Photon Screen, you don't think of the smaller Screen as primary and the larger universe as an extension; you think of the oval as the *visualized portion* of the larger universal reality; the oval of light catches only a tiny small portion of the whole world.

. . .

Then, just to complicate things, your neighbor opens his door and lets the dog out. The dog barks and runs toward a tree in the front yard. You see a squirrel in the tree. Then the dog sniffs the air and looks toward the cake. Does he share the dimensional universe with you and your neighbor? In a way, he does. He alerts you to the squirrel in the tree, and you can tell he smells and sees the cake. He lives in the same or a similar spacetime universe with you but does little to construct it because he provides little dimensional information. When he barks and gestures, he causes you to look where he is looking. He may, therefore, "have" an oval of a sort, but his information (his barking) does not describe the image he is seeing or smelling and does not tell you where it is. You cannot "see through his eyes" by listening to him bark. Only through hearing human language can you "see through your neighbor's eyes" to know where something is and what it looks like.

It is fairly easy to construct a dimensional universe with your neighbor, with or without his dog, because you are not moving relative to one another. You are in the same reference frame. You construct observational dimensions by factoring out space differences between observers without worrying about time. You take the finite space and time dimensions of the oval you see on the Photon Screen, extend them infinitely in all directions, and locate the things he says he sees on those extended dimensions. You create a sort of spacetime **Box** *that everything real is in. Because it is infinite in all directions, you appear to be in the Box, and because it is always there, it appears to exist independently of consciousness.*

. . .

Within the oval, the identity of space with time is not noticeable. 300,000,000 is such a huge number of meters for each second of time that you assume it is infinite when you use it to construct a rigid spacetime Box that contains you, your neighbor, and everything physical. You are unaware in everyday experience that c is finite, because it is so large a number that it makes space and time distinct for everything you see on the Photon Screen. Spacetime looks rigid and external in the room: the universe seems without relativistic dilations and quantum leaps. But when you encounter velocities near c, the universe's imperfections begin to show — the identity of space and time strains the structure of the Screen. Dilations show.

Knowing that spacetime is not rigid and breaks down at dimensional extremes, we try, still, to think of externally existing objects "causing" us to perceive. We remain grounded in a material world, ignoring talk of realms and rotations, thinking as we have always thought. Moving consciousness outside of space and time is as arduous as moving the Earth into orbit around the sun. It is not worth doing until it must be done.

View in Four Dimensions

As you move about the room, the time dimension rotates into space. As you interchange dimensions, you become the axis of rotation. You look at the vase, the flowers, the table, the record player, and your friend from a variety of angles, moving the oval of vision from one place to another. Remembering what you saw a minute before, your mind constructs a spatial framework made of time. Space dimensions extend in all directions beyond what you see.

Later, as you sit quietly in the room, nothing moves. The record player sits on the shelf where it always is; the flowers stay in the vase; and the lamp rests on the table. Even your friend sits still in her chair on the other side of the table, saying nothing. The past seems to move effortlessly into the future. Then you notice that the lamp has moved. You remember it yesterday in the middle of the table; now it is closer to the edge. And the floor needs sweeping, again.

· · ·

The traditional four-dimensional world is of matter in motion. From within the dimensions, consciousness appears to be a complexity of mechanical process inside of living bodies, one of which is you.

The alternative model presented here is the Photon Screen—the spacetime oval of visual consciousness. But how to envision vision? Why can we not hold a steady view of the Photon Screen as we look at it? Why is it not like a chair or a table?

There are spaces between chairs and tables; spaces that allow us to distinguish one image from another. But the Screen itself is spacetime *as a whole* and not an image in spacetime. You cannot hold it still, like an image in time. Space is not in space.

You can poke things with a stick, but you cannot poke the stick with the stick.

From the perspective of time, your experience of the Screen is a glimpse—an experience of all time and no time, from outside of time.

We see images in four dimensions of space and in time; we do not see their mass, but know some are heavier than others. With mass, *images* become *objects*.

Is *mass* a measure of matter in each object?

Or something else?

6

Dimension Five

Life is a daydream, with a body.

Spacetime-mass is an infinite set of points in five dimensions.

A point in spacetime-mass is a *quantum*.

A physical object is a finite range of quanta with length, width, depth, duration, and mass.

A physical object is an image in five dimensions.

Mass is the second time dimension, manifest in resistance to acceleration.

Mass exists at "right angles" to space and time, *foreshortened* in spacetime as seconds *per second*.

Mass is potential tactile perception.

Actual tactile perception is the body.

An accelerating object is a finite cross-section of space, time, and mass.

. . .

When you shoot the rifle at a target in the earth's gravitational field there is a second time dimension to consider, a dimension foreshortened in the vertical space dimension. The bullet will be accelerated towards the center of the earth at a constant rate of 9.8 meters per second per second. That is, the bullet will "fall" toward the ground as it speeds toward the target. Gravitational mass is not beyond, within, or outside spacetime — it is at "right angles" to spacetime.

You **feel** *mass when you accelerate, or when stationary in a gravitational field, standing on the floor, for instance. You* **see** *mass in the resistance of physical objects to acceleration, and in the downward acceleration of all objects in the curved spacetime of the earth's gravitational field. To offset the curvature, you must aim above the target.*

. . .

The path of a moving object is curved in the curved spacetime of a gravitational field. While this curvature is noticeable near an extremely massive object like the earth, a star, a galaxy or black hole, it curves very slightly around all objects, no matter how slight their mass. Light, which defines a line in space, curves in a gravitational field. The curvature of light near a star or black hole is a straight line through curved space.

Light determines a line in *space*, not in spacetime. Because light always travels at *c*, infinite dilation at the speed of light brings time to a complete stop. A photon never exists in time; light, therefore, is always now. Objects with mass, like the bullet, cannot reach *c*, and always have a time component to their motion. The bullet, therefore, travels a path in curved space*time*, a path of far greater curvature than that of light. A stone you drop to the floor is at rest in curved spacetime as it travels to the floor. As you watch the stone fall, the floor is accelerating up against gravity, a relative motion that makes the stone appear to be accelerating down.

Mass is a finite dimension of a physical object's Inertia. It is measured in terms of resistance to acceleration when subject to a known force. An object's identity is determined by its finite dimensions in length, width, depth, time, and mass. It is perceivable in five infinite dimensions or perceptual realms.

Without mass, objects would be images in spacetime that pass through one another without interacting. Momentum (mass x space/time) is conserved when objects collide without permanently changing identity; energy (mass x space^2/time^2) is always conserved, even as objects collide and change identity.

The tactile realm of perception is the body. We do not feel the body in space or in time; we only feel it in the second time dimension: under *acceleration* (or partial acceleration).

Dimension five is apparent in spacetime as a foreshortening into an existing dimension much the way a third space dimension is foreshortened in a painting or photograph. With foreshortening, the object does not appear in the foreshortened dimension; the dimension appears in the object. We do not see mass in an object; we see mass in how the object moves, that is, at more than one point in time.

A physical object is a five-dimensional range of points in space, time, and mass. Macroscopic physical measurements are made in these five dimensions — length and width are measured as area, motion is measured through time, density is mass divided by volume, momentum is mass times velocity, etc. In everyday macroscopic experience, points within the range appear to be continuous, touching one another in all directions. Objects seem to move smoothly from one location to another. But they do not. Space is coarse-grained - very finely coarse-grained. As an object moves across space it "quantum leaps" from one set of points to another — the pattern moves, not the points — like an image moving across a TV or computer screen. The points are extremely small, invisibly small, but they do not touch one another. Like the points on a pixel screen, spacetime is not continuous. Extremely small objects — objects smaller than points on the Screen — cannot be located precisely in spacetime-mass. A range of points cannot be smaller than a point, as a forest cannot be smaller than a tree.

. . .

You are back in the room with your friend. This time there is a switch on the wall that turns gravity off or on. You turn it off. You want to get an idea of what the mass dimension is like without gravity. Objects begin floating around the room: the record player lifts up off the shelf; the lamp floats over the table; the vase is no longer held down, and the flowers drift out of the vase. You and your friend are not held in your chairs. When you push the lamp, it moves away from you, and you from it, each accelerating during the push according to its mass, then continuing at a constant velocity across the room. The same happens when you push the record player or touch the flowers. Everything accelerates at a constant rate according to its mass. In the absence of gravity, objects are not held against a supporting surface and not subject to friction. Their identity is not confined by a gravitating body, and they are free to move independently of one another according to the laws of physics. A body in motion remains in constant motion until a force is applied. During a collision between two objects, forces are applied, and each accelerates away from the other at a constant rate. After the collision, each continues at a new constant velocity.

You notice that all velocities are constant except during collisions, and that all accelerations are constant even during collisions. No objects in the room are able to change their acceleration, in strength or direction. No objects in the room are able to control their motion — except for you and your friend. Only living objects — observers — are capable of non-constant acceleration. They move as if they are alive.

When you switch the gravity back on, it "pulls" flowers back into the vase and the lamp back onto the table. You and your friend fall back into your chairs. Because the two of you are not flying through interstellar space at millions of miles per hour, your meter sticks, clocks, and balance scales all agree. You are in the same reference frame, and everything is normal. There are no dilations.

But something else bothers you. Ever since you got back from flying through interstellar space at half the speed of light, you have noticed that your friend's watch is ticking at the same rate as yours, but it is **an hour behind**. *Your watch says four o'clock and hers, three. Your elapsed time in interstellar space was a full hour more than hers! If you happened to be exactly the same age as she at takeoff, she would now be an hour younger than you. This makes some sense, in that you saw her clock ticking more slowly than yours as she passed by at millions of miles per hour; but then again, she saw* **yours** *running more slowly than hers. What is the difference, then, between her reference frame and yours? How did she end up with less elapsed time? Why she and not you? Both reference frames were equal; neither was favored in any way, but now, there is an absolute (non-relative) difference between them. What's going on?*

You remember that as she whizzed by, you were sitting still. But from her reference frame, you were the one whizzing by. Velocity was purely relative. The only difference between the two reference frames was in how each came to be at a relative velocity. Somebody had to accelerate. You remember now that it was she. She was the one who stepped on the gas, circled back around, and passed by you at ½ c. That was the difference — the only difference — between reference frames. You never accelerated. That is why she is the one who experienced less total elapsed time. If she had sat still while you accelerated, you would be an hour younger than she.

*And your friend reports something else interesting: while she was accelerating, during the speed up and during the curving around, she reported a strong kinesthetic sensation throughout her entire body (a "g" force). Her entire tactile realm of perception was activated in precise proportion to the magnitude of the second time dimension (the per second **per second**) of her acceleration. While you were waiting for your friend to circle back you did not feel the sensation at all.*

. . .

Velocity and acceleration are not the same. They differ by the second time dimension. Velocity is meters per second and acceleration is meters per second per second. Acceleration is a change in velocity (speed or direction) but velocity already has a time component, so another time component is needed. Acceleration is not just more speed; it is an entirely distinct form of motion. Your friend does not *feel* velocity; she feels acceleration. If she is moving uniformly at 10 miles per hour or 10 million miles per hour, she does not feel her motion through space. What she does feel is *change* in motion. She feels only the *change* in the direction of her velocity as she goes over bumps and around curves, and the *change* in speed as she accelerates from, and decelerates back to, your reference frame. She feels the second time dimension throughout her body in the smallest bump, the slightest turn, or the most insignificant speedup or slowdown, but does not feel the speed at all, no matter how fast it may be. If the ride were perfectly straight and smooth, she would have felt nothing at all between the initial acceleration and final deceleration: she would have felt as if she were standing still. If there are no windows in the vehicle, she would have had no way to tell that she was moving once she reached a constant velocity.

Dilation in the rate of time is due to velocity.
Dilation in elapsed time is due to acceleration.

An *inertial* frame of reference is a spacetime frame at rest or in constant velocity (without acceleration). Dilations in space, time, and mass occur at high velocities between inertial reference frames. Changing from one reference frame to another requires acceleration — your friend had to move in relation to spacetime as a *whole*, and she felt it throughout her body — her tactile realm of consciousness. Under acceleration, she dipped into an additional dimension and stayed there until she returned to a constant velocity. During acceleration her reference frame was not inertial; therefore, her motion was not relative, but *absolute* in relation to the whole of spacetime. After she stopped accelerating and returned to a constant velocity, she was once again in an inertial reference frame at a high velocity relative to you. The passage of time in the first time dimension was relative: she saw *your* clock running slow and you saw *hers* running slow. If you had sped up to her reference frame at this point, going through all the loops and curves she went through to get there, your reference frame would have gone into the second time dimension the way hers did, and you would both be back in the same reference frame. Your clocks would be ticking at the same rate. But if instead, she decelerates back to your reference frame (that you never left), your clocks would tick at the same rate, but hers would show less elapsed time.

. . .

In the room, you move about by interchanging space and time dimensions. As you walk toward the vase or the record player you rotate time into space. But a simple time-for-space interchange describes only constant velocity. To change velocity—to move from zero to any constant speed—you need to accelerate; you need another dimension. It is that dimension you feel in your body as you begin to walk. Once in motion your overall velocity may be close to constant; but in the real world, unevenness in your step and in the floor, and passage around the table and other furniture add a continuing stream of vertical, horizontal, and forward-backward partial accelerations. Perfectly smooth constant velocity rarely exists.

You are always dipping into the second time dimension as you walk, ride, or fly, changing the direction or speed of your motion. You feel the g force either way.

When you arrive at the other side of the room, you reach your hand out to where you see or hear objects, knowing that there is potential tactile at exactly that location in time and space. Tactile sensation is waiting where you see objects.

. . .

The body is actual tactile perception. You rarely feel the body as a whole. What you do feel in everyday life are arms and legs and fingers and toes: parts of the body, and you feel them as partial accelerations. The body as a whole usually does not move when you touch things—partial accelerations balance out and you stay put. A partial acceleration in your finger is compensated by a partial (mostly unnoticed) acceleration in your toe—in the opposite direction. That keeps you from falling over backward when you touch the vase. When partial accelerations do not balance, the whole body accelerates, the entire tactile potential activates as you feel the "g" force.

But are we *in* the second time dimension the way we are *in* space and time? We are in the second time dimension to the extent that we are in the body. That seems obvious. The real question is, "What is the relation of the body to the other four dimensions?" Space is unique in showing a wide variety of objects in a wide variety of locations, all at the same time. We see dimensional relations within objects and in relation to each other, in how they are moving or not moving, and in how they interact. This reveals potential perceptual experience instantaneously.

The second time dimension is different. The location of physical objects in this dimension is not readily apparent; you cannot see the whole potential at any point in time; it must be constructed from at least three points in spacetime — three events — two to determine the object's velocity, and at least a third to determine its change in velocity. You can also know an object's location in mass, less accurately but more immediately, by *feeling* it. If you kick a bowling ball, it will accelerate more slowly than if you kick a soccer ball. Your toes will feel the bowling ball's resistance to acceleration more than that of the soccer ball. (More commonly, you would hold each ball up in the curved spacetime of the Earth's gravitational field to feel its "weight.")

Objects in spacetime-mass are potentially perceived in all realms, whether or not actually perceived. Perceptual potentials in the form of dimensions are coordinated with one another so that where you see an object is where you touch and taste it. Objects appear, therefore, to exist independently of consciousness, as if waiting to be perceived. There is no proof that they do or do not exist independently of consciousness, but potential perception gives them the appearance of independent existence. *Material substance* is potential perception in all five realms.

Material substance is potential perception.

An object's identity is determined by the contiguity of its range of points in space, time, and mass. It is less defined if there are gaps. If the shape of the object warps beyond recognition, it loses identity; if it disintegrates altogether, it has no identity. Objects often deform or disintegrate under acceleration; in fact, it is only under acceleration that object identity changes.

If there were no mass dimension—no second time dimension—physical objects would move through space but never change intrinsically. There would be no acceleration to distort, distend, create, explode, obliterate, or change one into another. There would be no way to combine smaller objects into larger objects. Everything would move at a constant velocity; one object would pass through another without colliding. Objects collide—interact with each other—only by maintaining contiguity in the second time dimension.

In an *elastic* collision, two objects collide and bounce off one another, retaining their original shapes. The range of points that constitutes each object dips into the mass dimension temporarily: object identity is distorted during the collision itself but returns immediately thereafter. Object identity is preserved, and momentum is conserved. The objects look the same before and after. If you calculate the mass times velocity of both objects (their momenta) and add them together, the result will be the same before and after the collision. Conservation of momentum means that if one of the objects is moving slower after the collision than before, the other object will be moving faster than before: momentum (mysteriously) moves from one object to the other.

Through an *inelastic* collision, however, an object is changed, created, obliterated, or altered in some way. The range of points before and after changes. Both objects dip into the mass dimension, but one or both do not return in the same form. Object identity is not preserved, and momentum is not conserved. (Only *energy* is conserved.) Some energy is transferred from the momentum of the two objects into heat energy — into greater motion of the points within the object.

Energy is the same before and after an inelastic collision — before and after a change in object identity — but momentum is not.

A billiard ball hitting another billiard ball is an elastic collision.

A billiard ball hitting a pillow is an inelastic collision.

The mass dimension as a whole shows up as a curvature of spacetime. We have known for over a hundred years that the paths of photons — defined as lines in space — are curved in the vicinity of massive objects like the sun. Spacetime is also curved in the vicinity of less massive objects like the earth. When you drop a pencil in the earth's gravitational field, the Earth and the pencil accelerate toward one another. You do not accelerate toward the earth when standing on the floor, because the floor supports you against gravity. The floor accelerates you up to counteract the downward acceleration of gravity; this is what you feel at the bottom of your feet and throughout the rest of your body. This is also what you feel in an airplane accelerating down the runway or in a car drag racing on a straightaway. You would feel this "g force" even if your body were accelerated where there is no gravity. Acceleration and gravity are the same thing—we have this from the same reliable source. If you were in a rocket ship beyond gravity but under acceleration, the mass dimension would scoot you back to the rear wall of the rocket and you would find yourself "standing" on that wall with your head pointed in the direction of the rocket's motion. Your body would feel as if you were standing upright in a gravitational field. There is no physical difference between the gravity and acceleration; they are both dimension five.

. . .

Now, back in the room, chatting with your friend, you are nowhere close to experiencing the high-speed dilations of space, time, and mass. This is the normal, macroscopic level of perceptual experience. The dilations show up only at dimensional extremes when you swoop past your friend at millions of miles per hour. None of us has had the opportunity to experience that kind of velocity in the real world, but the more we know about physics and astronomy—and consciousness—the more we realize that dimensional extremes are just as real, hard, and concrete as the everyday macroscopic level. We know through experiment—through science—that people would actually experience the dilations and other strange phenomena under conditions described in scientifically rigorous thought experiments. And we know that even if no one ever gets to fly at half the speed of light, different observers will have dimensionally different experiences of the same physical phenomena, depending on their reference frame. The fact that we need to reconcile what two observers see with a transformation factor proves that there is a physical difference between perceptual and observational consciousness. As soon as you say something about what you see, you become an observer, but you need to reconcile your dimensional perspective—both spatial location and reference frame—with that of other observers to come up with something everybody can call physical reality. Even sitting in the room, you experience this when you reconcile what you see with what your friend says she sees from the other side of the table.

· · ·

The differences between perceptual and observational consciousness are quantitative and qualitative. Quantitative, in that several billion people can tell you a lot more about what is going on in the world than you will ever see for yourself; and qualitative, in that perception comes to you in the form of photons and air molecules, while observation comes in words and numbers. (Words and numbers also come as photons and air molecules but are distinguished from normal sights and sounds by the informational order in which they appear. More on this later.) Perception and observation have been so closely aligned for so long that we do not distinguish between them. We have always thought that you and I see the same things because we were looking at the same things, "obviously."

Observation is a separate level of consciousness: let us call it, if we may, a "higher" level than perception. Not everything you experience is observation. Observation consists of dimensionally structured *potential perception*. It has to be what anybody can see, and it has to be described in a dimensional context. A story, a joke, a lie, or an exaggeration is not observation. Imagination is not observation. Spiritual experience is not observation. Only an actual perceptual experience communicated in a spacetime setting qualifies as observational. If I sit where you are sitting, I must be able to see what you say you see, or it is not observation. If you are deceived somehow, hallucinating, stretching the truth, or in the midst of a religious epiphany—if I cannot see what you see for any reason—you are not an observer. Or I'm not an observer if there is something I can't see that everyone else sees. We might argue back and forth all night as to what is really "there," or we might each decide to publish an article and let our peers determine what is truly observational and what not. But usually, we trust what observers say and save ourselves the trouble of verification. I have never seen the temples of Angkor Wat in Cambodia, for instance, but I have talked to people who have, and I trust them. I believe that if I were to go to where they were, I would see what they say they saw. So, if I am satisfied with the words and numbers of observational experience and do not require the more impactful and detailed experience of direct perception, I don't have to go. I can save a lot of trouble and expense by relying on observation. Sitting in a chair reading a book, talking on the phone, or watching television, I can know immensely more about the world than by trying to see everything for myself.

Observational consciousness is not entirely distinct from perception. An analogy would be the forest and the trees. One or two or twenty trees is not a forest, but a thousand is. The level of "forestness" does not arise at any special number of trees and is not in or around any particular tree. It exists as a wholeness over and above "treeness"—at a "higher" level. Close up you see trees; from a distance you see forest. From an intermediate distance you see both levels. As your friend describes what she sees in the room—or what she saw years ago at Angkor Wat, you experience both perception and observation. The images in your mind are observational, yet you hear the sound of her words in the auditory realm of perception. The "forest" is observation; the "trees" are auditory.

Throughout our waking hours we experience both the observational and perceptual levels. Observation is reducible to perception: your friend's experience of Angkor

Wat is reducible to the sounds coming from her mouth. The words are "nothing but" sounds, the way a building is nothing but bricks, or a forest nothing but trees. Order is eliminated when the forest or the building is reduced to nothing but its parts. Images words create in the mind are more than the sounds of which they consist. The order in which the sounds are arranged creates a level of reality over and above simple auditory perception; order creates a higher level of consciousness.

Above perception is a higher level of consciousness: observation.

Below perception is the level of cellular or quantum consciousness.

We do not ordinarily designate individual cells as conscious, but all that we see, hear, smell, taste, and feel is what our sensory cells experience. Nerves lead from cochlear, retinal, and muscular cells in ears, eyes, legs, and fingers to processing centers in the brain. Perceptual consciousness is reducible to cellular experience but exists over and above what cells experience themselves. The multicellular experience familiar to you and me is a composite of cellular experience arranged in a dimensional pattern. The sensation of an armrest is an orderly array of signals in the cells of your left elbow, the vase across the room an orderly pattern of experience in your retinal cells, and music from the record player an orderly arrangement of cellular experience in your cochlea. Order creates the picture within the dimensional potential.

Like the intermediate position between perceptual and observational levels, there is an intermediate position between macroscopic and cellular levels of consciousness. If you sit quietly, you can feel large numbers of "subtle sensations" (kalapas) in your elbow forming an image of "elbow" in your mind; you can hear the sound of air molecules without resolving them into an auditory image; and, as we have seen, you can see tiny dots of cellular experience not yet ordered into light when you look carefully at your retina. These are all difficult to experience and often require training in relaxation to appreciate fully, but you may catch a glimpse (not in time) at any time.

I have not spoken with any of my cells lately—they are not observers—but it seems likely that they share at least two realms of consciousness with multicellular organisms: the chemical and the tactile (taste and touch). Chemicals allowed through the cellular membrane are metabolized or absorbed into the endoplasmic mix and assimilated by the cell body. A cell must distinguish between beneficial and harmful molecules and know with what it is combining itself before allowing a molecule through its membrane. This is the most fundamental of sensory functions. On the multicellular level, we experience this realm at points where we allow external molecules into our cells, that is, along the digestive tract. The cells most sensitive to chemical sensation are those at the beginning of the tract: in the taste buds. It is here (and throughout the alimentary canal) that multicellular organisms share direct chemical experience with the interior of their cells.

Before cells absorb molecules, they touch them. They develop a sense of *potential taste*, that is, of how a molecule or conglomeration of molecules would be assimilated before it is actually assimilated. They learn what the object feels like on the outside of the membrane. The sense of touch alongside the sense of taste creates a dichotomy between the inside of the membrane and the outside, between self and world, between subject and object. The duality requires a clear distinction between tactile and chemical information: a separate information potential on the cellular level. On the multicellular level, we understand this duality in terms of space within and space without, but it is doubtful that cells have a sense of space. There is only the qualitative difference between the tactile and the chemical, as far as we can tell. But it is important to note that though these separate realms are distinct on the cellular level, the tactile is reducible to the chemical. The tactile realm consists of shifting concentrations of potassium ions on the cell membrane. When an external object touches the membrane, potassium ions rush to the scene. To distinguish between molecules on the inside and outside of the membrane (which is, of course, essential to its viability) the cell, whatever the dynamic of its "consciousness," must sense a qualitative distinction between the motion of potassium ions and that of other chemicals. The cell creates a wall between these two realms of perception that evolves into the dualistic experience of multicellular organisms: the self and the world, subject and object. The chemical realm is within; smelling, seeing, and hearing are without; the tactile realm is in between.

Smelling, seeing, and hearing—the spatial realms—are each reducible to touch and taste, that is, to the cellular level. The olfactory realm consists of cells in your nose tasting airborne molecules. These cells do not have to be in contact with the *object* of olfactory perception. The object—the thing you are smelling—is distinct from the smelling itself, and out in space somewhere beyond the body. The auditory realm is reducible to minute tactile sensations of cochlear cells in the inner ear, and the visual realm to even tinier tactile sensations of retinal cells. Cellular sensation in any of these three realms becomes perceptual experience of the organism as a whole, in space. Information channels provided by space dimensions create an entirely new level of meaning—and of being—beyond the tactile and chemical experience of sensory cells. Space dimensions are unique to multicellular (animal) consciousness.

That there is a clear dimensional distinction between auditory and visual information is indicated by auditory information coming in the form of *longitudinal* waves—waves in the direction of propagation, while light comes in the form of *transverse* waves—at right angles to the line of propagation. Each realm has its own dimension, but once coordinated into spacetime-mass, dimensions become interchangeable.

Space dimensions do not exist at the cellular level of consciousness. (They exist at the level of *cells* but not at the level of cellular *consciousness*.) Retinal cells, for instance, sense the touch of photons without perceiving the spatial pattern of a physical object (the range of points). At the cellular level all sensory information is quantum, and space, time, and mass dimensions dissolve into one another. Momentum, which is mass times space divided by time (kg–m/sec) on the macroscopic level, has no separate dimensional components on the quantum level. Because dimensions are not separate, we cannot know an extremely small particle's location in space and time if we know its momentum. The Screen is too coarse.

Velocity, momentum, and energy are more fundamental than their dimensional components. Extremely small images slip between the quantum points on the Screen.

This is amazing to consider. This is something built into the fabric of the universe — something that has nothing to do with limitations in our measuring instruments. It might seem common sensical that you could divide space, time, and mass any number of times and get smaller and smaller pieces, but you cannot. There's a built-in limit to how small something can get and still be in spacetime-mass. Here's the equation:

$$\Delta mv\Delta x = h/4\pi$$

The *h* is the Planck Constant, an extremely small number (6.62×10^{-34} kg-m2/sec), here divided by 4 x 3.14.

What the equations means, very simply, is that uncertainty in momentum (Δmv) times uncertainty in location (Δx) is an extremely small number; and the smaller one of them is, the larger the other has to be. For a very small particle, the more you know about momentum, the less you *can possibly* know about location. There is no getting away from the uncertainty of space, time, and mass at extremely small magnitudes. That's how the universe is made. It's not smooth and continuous; it's grainy — or "quantized." Multicellular consciousness is a composite of cellular consciousness that smooths out the graininess only on the macroscopic level.

Points in spacetime-mass do not quite touch, and tiny particles get lost between points. Dimensions dwindle out of existence when objects are too small to constitute a *range* of points. An object smaller than the touch of a retinal cell cannot be perceived in spacetime-mass because spacetime-mass is too coarse. On the cellular level, there is no distinction between tactile and visual consciousness, and therefore no distinct dimension. Space, time, and mass fall apart.

At the quantum level all is reducible to touch and taste. No space, no objects, no ranges of points: no forest, only trees. The forest disappears into cracks between the trees.

Dimensions exist to make multicellular sense of cellular experience.

Cellular experience on the quantum level makes no sense to the multicellular mind.

. . .

Let us look first at an object on the multicellular (macroscopic) level. In the center of the room where you are sitting with your friend there is a perfect cube: 10" x 10" x 10" in three finite dimensions of space. It is moving across the room from left to right at 2 meters per second, and because it is unsupported in the Earth's gravitational field, it is falling at 9.8 meters per second, per second. The edges of the cube are perfectly straight, the corners are sharp, and the sides are perfectly square. No matter how closely you look at the cube, its location is perfectly defined. You find a magnifying glass and take a closer look at one of the edges: it is still perfectly straight. Then you find a high-powered microscope and look again: still perfect. It seems the shape of the cube, and of the space it is in, is going to be perfectly smooth and continuous no matter how many times you divide space into smaller and smaller pieces, but you wonder if there might be a limit. You have taken your investigation as far as you, yourself, can see so you call in a team of scientists with electron microscopes, cloud chambers, and particle accelerators to extend what they are able to see. You don't know how to use their machines, but you trust what they say because they check each other's findings. You do not believe any one of them until the others arrive at the same results.

One of the scientists notices, and the others agree, that subatomic particles seem to dance back and forth on either side of an invisible line that vaguely defines the edge of the cube. Sometimes they appear inside the line and sometimes out; at other times they do not appear at all. At an instant in time, a single particle cannot be found at any distinct location in space — but it is more likely to exist in some locations than in others. The particle is more likely to be found close to the line, but occasionally ends up way outside or way inside of the line. The closer you look, the less the defined the edge line becomes, eventually devolving into something of an "average" location for particles, though few individual particles actually occupy it. The object you are calling a "particle" is more like a wave spread out in space than like a point in space, yet when the particle is observed, the wave collapses and the particle occupies a single point — a point **in the past**. *In the present you can't see where it is, only where it probably is, and when you do see where it is, you know nothing about what it its doing. Points in space, in time, and in mass, are not quite touching — do not smoothly connect with one another. A subatomic particle appears to leap between one point and another, but there is no space between the points it leaps between. Furthermore, the scientists report that the smaller the particle's mass, the larger its wave — the more widely spread it is in space. The "fabric" of the physical universe is an interweaving of space, time, and mass that unravels at the quantum level.*

The scientists also report that when they look at a large number of subatomic particles from a little more distance, the edge of the cube reappears. Some particles are observed inside and others out, but the average location reveals an edge line that defines the cube as a macroscopic object. It is still impossible to tell where a **single** *particle is between observations, but the* **aggregate** *behavior of a large number of particles is predictable. The probability of a single particle's* *behavior on the quantum level becomes the predictability of aggregate behavior on the macroscopic level.*

. . .

Here's an equation that summarizes what the scientists find out:

$$\lambda = h/mv$$

Again, it looks scary, but is actually quite simple once you get over the Greek letter. The λ is the wavelength of a particle, or roughly the size of the area in which it might exist. The greater the wavelength, the more fuzziness, or uncertainty, in the particle's location at any moment. The h is the Planck constant again, an extremely small number, but one that never changes. It pops up regularly in quantum physics. m is the mass of the particle, and v is the velocity of the particle. What the equation means is that a particle's degree of fuzziness in spacetime-mass is going to be extremely small (because the h is small), and that the smaller the mass and velocity of the particle, the greater the fuzziness. A bigger particle with more mass and velocity is less fuzzy and has a more certain location. There is virtually no fuzziness at all, therefore, in the location of the cube (a really big "particle"), only in the particles making up the cube. Zooming out from the quantum level, the multicellular level of consciousness takes over from the cellular, and the edges of the cube become clear. (This equation says what $\Delta mv\Delta x = h/4\pi$ says, but in a different way: instead of the Δ expressing uncertainty, the λ expresses a wave.)

The coordination of all dimensions, and all perceptual realms, into a single, multi-dimensional universe is the construction of hearing, smelling, tasting, and touching into a single spacetime structure based on seeing—on light. c is the spacetime structure of perceptual consciousness as a whole. Information potentials for all perceptual realms are transformed into that of the visual. The universal role of light is manifest by the fact that c, the "speed" of light, is the same *relative to all objects in the universe* regardless of their velocity relative to each other.

To illustrate this feature of the universe we return to interstellar space:

. . .

You and your friend are at rest in the same inertial reference frame (not moving in relation to each other). She shoots a bullet that you measure to be traveling at 500 miles per hour. She measures 500 miles per hour also. But then you accelerate to 100 miles per hour, changing your reference frame. You turn, swoop around her, and as you pass by you shoot a bullet in the direction of your motion. (It doesn't matter which one of you accelerated for this experiment—it could have been her.)

You measure the speed of your bullet at 500 miles per hour while she measures it to be 600 miles per hour, as expected. The velocities add up. Now it's her turn. She accelerates to 0.5 c, circles back, and as she passes by you, shines a flashlight in the direction of her motion. She measures the speed of the light coming from the flashlight at c, as expected. You measure the speed of the light from her flashlight, too, expecting it to be 1.5c, **but it's not**. *You measure the light speed to be c, the same as she does! The two velocities do not add up! Despite the relative velocity of the two reference frames, you and she both measure the speed of light to be the same. Light always travels at c relative to all reference frames! If light could travel as fast as it wanted through empty space, and at different velocities in relation to different reference frames, there would be an externally existing "objective" space that light is in. But there is no such external space. The visual realm defines spacetime within vision and also provides a universal template for all perceptual consciousness.*

· · ·

Light is visual consciousness.

Spacetime is in light.

The dimensional structure of vision is the dimensional structure of the universe.

Model of Dimension Five: The Quantum Screen

Space dimensions are the transition between cellular and multicellular consciousness. They are created to compartmentalize and coordinate perceptual information. When cells combine into multicellular organisms, some become specialized sensory cells that communicate with other cells in the body. Signals from five separate information channels are integrated into a multidimensional matrix that becomes the perceptual experience of the organism as a whole. Where space and time are the essence of the 4d Photon Screen, space, time, *and mass* are the essence of the 5d *Quantum Screen*, the picture in which all sensory information is arranged.

We are not interested here in the cell as a protoplasmic enclosure; we are interested in cellular *signals* that aggregate to become multicellular perception. The cellular signal — not the anatomic cell in space — is the primal reality in this understanding of perceptual consciousness.

A quantum is the fundamental unit of perceptual consciousness: the smallest possible unit of energy and the signal between a single cell and the multicellular organism. Photons are quanta of a variety of frequencies, each frequency perceived as a separate color. The per-second of a particular photon's frequency becomes the per-second of mass when aggregated with other quanta into a macroscopic object. (A photon has no mass until it is aggregated with other quanta.) The Photon Screen, then, is what we see on the 4d "surface" of the Quantum Screen. The 5d "depth" of the Screen is the frequency of the photons, or, in their aggregate, the mass of a physical object. Mass is the "glue" that holds a quantum pattern together as a 5d image.

The Quantum Screen presents a non-visual dimension to perception as a whole, but a more complete model of perception than the Photon Screen for a variety of reasons. On the macroscopic level, the Quantum Screen shows an additional depth in the behavior of physical objects in the form of resistance to acceleration. It also shows the mass dimension as a whole (and its correspondence to the tactile realm) in the bodily g force felt under acceleration. In relativity, the Quantum Screen shows its limitations near extremely massive bodies in the gravitational curvature of 4d spacetime into mass and shows the distortion of images at extreme velocities as a space and time rotation into mass. On the quantum level, the reduction of multicellular to cellular sensation is shown on the Quantum Screen model in the disintegration of space dimensions, and the composite structure of multicellular perception is shown in the granular structure of the Screen.

The Photon Screen is what we see as we look around the room, but not what we feel. If you are able to feel actual tactile perception in your body as you walk about the room, and also see objects you are not touching in the room as locations of *potential* tactile sensation, you are experiencing the Quantum Screen. Potential tactile experience is the feature of the Quantum Screen that encourages us to look both ways before crossing the street.

The Photon Screen—visual consciousness—exists only on the macroscopic level. The Quantum Screen is the foundation of the Photon Screen, reaching below the quantum level and presenting a much wider view of physical reality.

If you are able to experience the Quantum Screen as you look around the room, the Photon Screen will be limited to those quanta actualized in the visual realm.

An individual quantum, as a point in five dimensions, has no intrinsic or finite dimensions. It has no shape, no size, no mass, no extension in space or time. It is not a *thing*, as such: not an object. But it is the stuff of which objects are constructed in space, time, and mass, or more precisely stated, a point in the framework within which objects are extended. As the quantum is structurally based on the photon, the framework of the Quantum Screen is structurally based on the framework of the Photon Screen. Objects on the Quantum Screen, unlike on the Photon Screen, are images with mass; they resist motion, interact, and you can touch them.

Quantum usually means a tiny "bundle" of actual energy, such as that of a photon impacting the retina or an electron impacting a photographic plate. But it also means tiny bundles of quantized spacetime through which subatomic particles "leap." (Electrons, for instance, are said to "leap" between quantum levels in their orbits around nuclei.) In this sense, a quantum is less actual than potential; it is more a tiny point of potential perception — a point on the Screen — than a bundle of actual energy.

It is difficult to envision a line of such quanta stretching infinitely in space, shoulder to shoulder, "touching" one another without touching at all. It is more difficult to envision an infinite *plane* of such quanta, and then *three full space dimensions* of these tiny balls "filling" all of the space in the universe. On top of this multiple infinitude, we must also envision dimensions of *time* and *mass*. But the five quantum dimensions, if you are able to envision them, are not what you see; they are where you see it. They are "real" only as a mental framework — place holders for actual information. Quanta in this sense are not themselves substantial and not *in* anything. They are the dimensional universe — perceptual experience as a whole.

Spacetime-mass below the quantum level is too grainy to properly locate subatomic phenomena. "Particles" smaller than quanta slip into cracks in the non-space between points on the Screen. They behave either as particles or as waves, but never both at the same time. When we try to understand them as tiny pieces of "matter," we cannot say where they are; and when we see where they are, we cannot say what they are doing. We observe a particle at a single point in spacetime and again at a later point, but we have no idea what the particle is, if it is, or if it is "the same" particle between the points that we see. There is no space that "the same" particle passes through.

All we see is the pattern, with gaps. The pattern comes of proximity. Closeness of points in space shows where a pattern begins and ends; closeness in time shows motion of a pattern through space; and closeness in mass shows the interaction of one pattern with another. If there are three or four observations at points in a line on the macroscopic level, you may assume that they constitute an object moving through space from one point to the next. You may also assume that if you were to make additional observations along the line you would find "the same" object. On the quantum level, however, there are no observations between quanta. There are no points between quanta and no physical connection between the points that can be connected into an object. The big question in quantum mechanics is, then, "What is happening between points of observation? What *thing* connects them?" The assumption behind the question is that there should be something "out there"—something external to observational consciousness—*causing* the connection that causes conscious experience. But, like the pixels on a computer screen, or the ink dots in a newspaper photograph, there is nothing of the image between its points.

If we allow a series of quantum observations to accumulate over time, a pattern develops. A stream of electrons on a photographic plate, for instance, allowed to pass through a small hole in a barrier, will after a while show a target-like wave pattern on the plate. A concentration of dark electron "hits" will develop at the center with concentric blank and dark rings alternating outward. One particle at a time lands in only one place: at the center or at one of the rings farther out. The concentric rings reveal that the stream of electrons has a wave pattern revealed over time. But even a single electron has a wave pattern: there is no way to know where it will land, but it can land in only one place of many, and therefore has a probability of landing at any particular place in the ring pattern. The *wave function* of the electron expresses the electron's potential to form a pattern. The pattern is revealed in the aggregate as an actual pattern in space and in each individual electron as a probability (the wave function squared) of where it will land. An electron looks like a particle at each isolated observation and looks like a wave between observational points. An individual "act of observation" can be anywhere in the probability wave, but the location cannot be known until the observation. This gives the illusion that the observation causes the electron to be at that location, a phenomenon known as *observer created reality*. Observation interacts with the wave function, causing it to "collapse" at a particular location. Once the electron is observed, the wave collapses because there is no probability of it being anywhere else.

The "act of observation" is, of course, consciousness. The idea of *observer created reality* requires an *interaction* of consciousness with a reality external to consciousness. This idea does introduce consciousness into physics, which is a step in the right direction, but still seeks ultimate reality beyond conscious experience, which does not exist as far as we can possibly know. Even as we come closer to a realization that there can be no experience beyond experience, we find ourselves still searching for an external reality. At some point we will have to be satisfied with experience itself—with what we actually see, hear, touch, taste, and smell—coordinated into a single picture, a picture that seems material due to the coordination. When dimensions are known to be structures internal to consciousness, we will no longer look for physical reality outside of consciousness. At that point we will begin to understand what physics has been trying to tell us. Observation does not create reality or cause consciousness; it *is* consciousness and *is* reality.

Science is potential perception—what anyone can see at any time under the same conditions: not emotion, imagination, opinion, or personal reflection. It is constructed by observers testing and publishing what they see. A scientific fact is an objectively provable fact. Scientists do not, therefore, trust subjective experience. What you see within the perceptual realms, or within your mind, is not science unless I can see it there too, and I cannot. Not directly. Neither of us sees the other's consciousness, so we build a structure of observational experience outside of either of us.

What follows below, a description of the Quantum Screen, is not science because it requires an acceptance of subjective experience. I justify the following exercise by a guess that, to some extent, you will "see within" what I see within, though I cannot point to it in space. I cannot go in there with you.

We begin with the Photon Screen, the oval-shaped field of visual consciousness. Look at it again with your eyes closed. It's still an oval, with patches of dark and light, but no resolved images. There is nothing specific to look at, but if you allow the mind to calm for a few minutes, you will begin to notice the tiny dots of light flashing on and off. There are thousands of them, maybe millions, very close together but not touching. They are so small that it is unclear where one ends and another begins. You can barely see just one dot; it is not an object "out there" in space, but neither is it imaginary. The flashing dots are retinal cells passing signals to multicellular consciousness.

With your eyes open, you are still looking at the Photon Screen, though the Screen itself is hidden by the images that light up on it. The patterns you see are photons in four dimensions. Each photon is the tactile sensation of a retinal cell, a point in spacetime, or an event. But each photon also has a frequency, seen as a color, measurable as a per second, or a value in the second time dimension. This is the mass dimension, the color of the photon you see is its tactile per-second value as it touches the retina. Retinal cells feel the frequency and send us the color. The dimensions of the Screen are structured in a way that coordinates the informational channel of vision with that of hearing, tasting, and smelling. But what about touching?

Restricting perceptual consciousness as a whole to the Photon Screen would limit us to a very small world. Everything in any realm would have to be in the oval. To create the Quantum Screen, the non-visual "view" is widened from the oval and extended infinitely in all directions. This is facilitated by moving your eyes, or your head, and extrapolating dimensional relations beyond the immediately visual. When you hear something behind you, or smell something in the next room, or touch something you cannot see, each object has a location on the Quantum Screen beyond the Photon Screen. But you will see it if you move the oval in the direction of the non-visual information. The spacetime extension of the Photon Screen beyond actual seeing creates potential visual information where and when you touch, taste, or hear. This creates the illusion of material substance throughout the universe.

The mass of objects shows up in spacetime as resistance to acceleration and in interactions among objects. Basketballs and bowling balls appear to "cling" to space as you push them, each according to its value in the mass dimension. Parts of your body show up on the Photon Screen; you see your hand in front of your face, or your knee as you sit in a chair. But how does the body become a pattern on the Screen? How do you become — or appear to become — an object in spacetime? The answer is key to understanding the difference between the Photon Screen and the Quantum Screen.

On the Photon Screen, retinal cells are flashing on and off within the oval, experiencing photons as tiny tactile sensations, reporting their experience to you, the multicellular mind. But what about non-retinal cells? Cochlear cells report tactile sensations as sound, and ordinary body cells, like those in your legs, fingers, neck, and toes report touch. How do you coordinate what they experience into the multicellular mind? Can you feel each tiny tactile pulse from a single cell in the body the way you see it in the oval? Probably not, but you can feel the patterns they create in the aggregate, and you do, in the form of "arms," "elbows," "buttocks," "jaws," and "shoulders." These are all patterns of tactile sensation reported by body cells. You experience them where and when you see parts of the body on the Photon Screen and your body becomes an outline of tactile sensation in spacetime. But because tactile sensation is a fifth sensory realm, and the Photon Screen is only four-dimensional, there is no room for tactile information. Objects in contact with the body (actual tactile information) create partial accelerations that the body "feels" in the second time dimension. Objects not in contact with the body show the tactile potential in how they accelerate and interact with other objects. This five-dimensional potential is the *Quantum Screen*: four dimensions of spacetime with mass foreshortened at "right angles." When the body as a whole moves in relation to spacetime as a whole, the entire tactile realm is activated in the form of the g force.

This extension of the Photon Screen to become the Quantum Screen is not of actual photons—not of actual vision. We could call the points that we do not see beyond the oval "potential photons," in that they become actual when you turn your head, but a better word for them is *quanta*. A photon is a type of quantum—the smallest possible unit of visual energy—and the only single quantum we experience. (Some highly disciplined meditators are capable of perceiving non-visual quanta, or kalapas, but this is exceptional. Most forms of energy consist of quanta in the aggregate.) A quantum is something of a generic version of a photon: an abstraction based on the dimensional framework of photons, without the light. A quantum or range of quanta may manifest in the form of kinetic energy, or of sound.

In using the dimensional structure of light beyond light itself, the multicellular mind coordinates potential vision with every other perceptual potential. Objects that you touch become smellable; objects you taste become visible. Actual perception in any one realm is potential perception in every realm — a coordination of sensory potentials that creates the universe.

Space is related to time on the Photon Screen by c, the structure of light, while mass is related to energy on the Quantum Screen by c^2, the square of the structure of light. Four dimensions of space and time are created as the tactile perception of individual retinal cells becomes the perceptual consciousness of a multicellular organism.

View in Five Dimensions

The physical world is now complete. All objects, systems of objects, and measurements on the macroscopic level are understood in terms of meters, seconds, and grams. Temperature is energy unorganized into patterns.

Matter may or may not exist, but it is never experienced directly and is not needed to explain perceptual experience. Consciousness is often forced into spacetime-mass as a mysterious emergent property of organic complexity, but it is not experienced as such. Objects fit into dimensions, but the experience of objects does not.

The *temporal* realms of consciousness (chemical and tactile) are experienced on both the cellular and multicellular levels. The three spatial realms, olfactory, auditory, and visual, are experienced only on the multicellular level. It may be more descriptive, in evolutionary terms, to call the chemical and tactile realms "Dimension One and Dimension Two," and call the multicellular realms "Dimensions Three, Four, and Five." But because they are interchangeable, there is no natural order.

The leap of consciousness from the cellular to the multicellular level is an evolution of *self*. Macroscopic consciousness arises when the aggregate behavior of quantum particles becomes physical reality—when the law of averages becomes a concrete physical object. The *multicellular self* arises when individual cells see collectively, and more importantly, when they are able to do things collectively.

Dimensions are fundamental to multicellular consciousness, but not fundamental to consciousness. Dimensional extremes are the edges of multicellular consciousness.

If consciousness is identified with self, solipsism becomes a necessity. If not, we will need another dimension.

. . .

You are sitting in your chair in the living room. The record player is off, the flowers are gone from the vase, and you are relaxing with your eyes closed.

You hear the oven door open in the kitchen.

The experience of hearing the springs expand in the oven door comes to you as **actual** *auditory information in the context of* **potential** *visual, tactile, chemical, and olfactory information. It sounds like someone is in the kitchen with an oven glove, opening the door, but you don't actually experience any of this. The kitchen and the oven do not exist in perceptual experience. They are "real" only as potential experience. The same is true of objects in other rooms of the house, and in houses down the street and across town.*

In a material universe, you are missing most—almost all—of what is happening in the world. There are kitchens and ovens on the other side of the planet and stars and galaxies in an ever-expanding spacetime universe of billions of light-years in every direction. In this universe you are but a tiny atom in a (nearly) infinite expanse of reality. This is the **Box***. The Box exists with or without your conscious experience. It existed before you were born and will exist when you are dead. You are a tiny morsel somewhere within. The Box is the* *"larger context" of your hearing a material oven door open in the next room.*

. . .

Your actual experience on the Quantum Screen is the same as in the Box. The only difference is the "larger context." The dimensional structure of the Screen (the spacetime context of what you heard in the next room) informs you that there is potential experience at that location for seeing, tasting, smelling, and touching. That's all. There is no universe out there waiting to be perceived. The potential experience is not actual, but real, due to its spacetime context. When you smell the cornbread a minute later, and taste it a few minutes after that, potential experience becomes actual. The perceptual realms, as information potentials, are proven to be dimensionally coordinated: no need to assume, without evidence, the existence of a vast external universe. You experience everything in the universe at every moment, actually or potentially — mostly potentially.

But the Quantum Screen is an enormous epistemological price to pay for a few physical oddities at the outer reaches of space and time. Why, you might ask, do we have to consider the strange behavior of physical objects at dimensional extremes when trying to understand everyday events in the kitchen? Why such elaborate metaphysical gymnastics for the sake of a distant "larger context?" Does the fact that light is not in space mean that we have to reconsider, and remodel, the entire physical universe?

We need the Quantum Screen the way we needed Copernicus 500 years ago. The only reason the Earth had to be put in motion — *the only reason* — was to explain the behavior of the other planets. Planetary motion was, and is, something hardly anyone ever notices. When do we ever see the planets? They are hidden in daylight for the vast majority of our waking hours, and usually hidden at night by clouds, buildings, and streetlights. When we do see them, most of us cannot distinguish them from stars. Even if we knew which were the planets and could see them all day and all night, their motion would be only in relation to the stars. Stars and planets and sun and moon all move across the sky together from east to west: to distinguish planetary motion, you have to view the sky night after night for weeks and months and take note of the change in planetary positions, not in relation to the horizon, but in relation to nearby stars. But even this does not require Copernicus. What puts the sun at the center is the planets moving in one direction across the sky *and then moving back the other way*. This does not make sense in an Earth-centered universe. (Some people suggested that angels were pushing the planets this way and that.) For the sole purpose of creating a "larger context" for everyday life, Copernicus undermined the very foundation of the metaphysical reality of his time, angels and all, by saying that the Earth is itself a planet. The outer planets appear to move backward at times because the Earth overtakes them in its orbit closer to the sun. Today, we "know" that the Earth moves around the Sun but rarely think of the metaphysical anguish our forebears experienced in abandoning the Earth-centered universe.

Quantum mechanics and relativity theory mean that the Box will have to go, despite the metaphysical anguish.

. . .

Then, a minute after you hear the oven door open, you hear a voice from the kitchen,

"The cornbread is ready."

*The Quantum Screen suddenly expands from five dimensions to six as you hear words that enable you to experience the physical world beyond perception—through someone else's experience. An entirely new dimension appears at right angles in every direction—**observational consciousness**. You know what is happening in the kitchen without being there.*

7

RECAP

Back to the original question: "Why can't an object travel as fast as it wants to through empty space?"

You may have noticed that in Dimension One through Dimension Four I use the word *image* to describe a concentrated pattern of perceptual experience, while in Dimension Five I switch to the word *object*. This is deliberate. An *image* is a pattern without substance. Through a telescope or projected onto a screen, for example, an image is a representation of something other than itself, not a material thing occupying the space where it is seen. An *object*, on the other hand, is really there. When you touch it, it touches you back, right where you see it. I use the word *object* for 5d because its visual image coordinates spatially with its tactile potential. It really is there, but really *there* only in the sense of the dimensional coordination. (Being *there* when you do not see it or touch it is not experienced and therefore unprovable.) So, by the word *object*, I mean a *five-dimensional image*: an image that, due to its fifth dimension, appears to have a material existence outside of perceptual consciousness because you can touch it.

The image appears on a space-time screen that looks perfect on the macroscopic level—so perfect we do not notice it.

An object cannot move as fast as it wants through empty space because it is an image on a screen that cannot show more than c meters in a given second.

On the retina, light waves emerge from light particles. The *spacetime* of visual consciousness emerges above and apart from the *mass* of minute tactile sensations to become the dimensional structure of the Photon Screen. This is the world we see.

But the Screen is limited by its structure.

As extreme space-for-time relations (velocities) of objects on the Screen approach the space-for-time limits of the Screen itself (c), *space* shortens and *time* slows with the reduction of spacetime back into *mass*.

Space and time appear to go on forever in straight, perfectly rectilinear lines in all directions. But the Screen is not perfect. It is structurally limited by the fact that space is reducible to time, and light reducible to the tactile sensations of cells in the retina. The Screen consists of cellular experience categorized and coordinated into a multicellular wholeness that transcends its separate parts, but that remains reducible to its separate parts. At dimensional extremes, finite dimensional relations within images approach the seemingly infinite dimensional relations of the Screen itself; the space-per-time velocity of a high speed image approaches the space-per-time structure of the Screen, so the image is distorted.

Seeing an image moving close to the speed of light is like looking at an image on a pixel screen from less than a perpendicular angle. As the angle of view approaches the horizontal, points on the screen come closer together. As space shortens, time slows, and mass increases, the Quantum Screen bulges out around the edges of the Photon Screen.

An object cannot move as fast as it wants through empty space because it is an image on a screen that cannot show more than c meters in a given second.

8

DIMENSION SIX

A perception is a thought in a dimension.

Dimension Six is order. It is finite order in the infinitude of entropy. Dimension Six is *practical thought*, *observation*, and the *future*; it is order that distinguishes life from non-life.

Entropy is an infinite set of points in six dimensions. A thought *that could be perceived* is a finite shape in six dimensions.

A living being is a finite cross-section of space, time, mass, and order.

Dimension Six is not free imagination, opinion, dreaming, or hallucination; nor is it seeing or touching. It is thought that *dimensionally coordinates* with perception: thought that is not actual perception but that could be actual perception. Practical thought adds a dimension — Dimension Six — to the physical world.

Practical thought is not perceiving objects where they are; it is thinking of them where they could be. It is hearing what an observer says he sees, thinking of the object where he describes it in space and time, and knowing that you could see it there yourself if you were to go and look. It is *potential* perception: potential because it coordinates with the five dimensions of perceptual consciousness. Dimension Six is information that allows you to "see through the eyes of observers." It is information *reducible* to the auditory, tactile, or visual realms, but ordered in a way that produces an additional realm of dimensional images that are potentially perceived wherever and whenever observed.

Dimension Six is thought that is *doable*—thought that is physically practical—capable of rotation into space and time—thought that *could become* physical. It is thought that enables placement of an object in a location it does not currently occupy. In Dimension Six we *do* things.

Entropy is the infinite potential of Dimension Six; orderly energy is its actual. It is the potential of disorder against which order—information—becomes possible. Within the entropy of Dimension Six is the tendency of living beings—observers—to overcome disorder to stay alive. The sounds an observer makes, the symbols he uses, the things he does, create a new realm of consciousness—and a dimension—over and above the five dimensions of the physical world.

The difference between perception and observation is often overlooked. Dimension Six often appears as thin as the third dimension of a sheet of paper. What you see is five-dimensional; what someone else says they see is six-dimensional, even though the object they are talking about has only five intrinsic physical dimensions.

Dimension Six is not in space-time-mass; it is an infinite realm of entropy foreshortened in spacetime-mass the way mass is foreshortened in spacetime. Where velocity is change in position, and acceleration is change in velocity, dimension six is change in acceleration. This sounds physically complex, but it is easy to see in everyday life: observers are distinguished from physical objects by the way they move. They change their motion in an orderly manner, seeking out objects they wish to contact and avoiding others in their path. Observers change motion or change the motion of objects with which they interact in a manner that creates the distinction between living and non-living. Observers accelerate in an orderly manner. Where Dimension Five is the dimension of physical reality, Dimension Six is the dimension of life.

Science is the deliberate creation of observational consciousness—shared information as to where and when objects may be perceived by anyone. Long before there was science, observation existed in the form of communicated words and numbers; now it is evolving rapidly beyond words and numbers. Through electronic media, present day humans are increasingly capable of seeing through the eyes (and cameras) of one another. Far more observational information is available than a generation ago, and a higher degree of collective consciousness is rapidly forming. Today we live in a world much larger than presented by the perceptual realms alone.

An understanding of Dimension Six is complicated by its evolving structure and also by its connection to a higher level of consciousness. Evolution to the higher level of collective consciousness requires a collective self or selves; it is analogous in many ways to the evolution of multicellular from cellular consciousness. Growing interdependence among cells millions of years ago, the diminution of their direct contact with the outside world, and the development of specialized sensory organs led to the five-dimensional physical world with which we are familiar. Similarly, in the present day, growth of human interdependence, decrease of direct sensory contact with the outside world, and evolution of electronic media, is leading to a higher level of collective consciousness among humans.

Dimension Six is an infinite potential of new information: "many worlds" at right angles to space, time, and mass. But this is not the "many worlds" interpretation of quantum mechanics. Dimension Six is this world — orthogonal, connected — not parallel. We see it within the mind as an infinitude of possibilities beyond the physical world. We collapse the non-physical to the physical when we do things, when we create science, when we see observers foreshortened in spacetime. Dimension Six arose in the human mind with the first spoken word and is evolving now into the "electronic retina" of the

Pixel Screen. Dimension Six is where we will grow, and how we will grow, into the space that is not space.

The enigmas of modern physics arise from assumptions — assumptions underlying dimensions, light, and observational consciousness.

Material substance is not experienced directly, and consciousness in others is not experienced directly. Neither is necessary to understand physics or everyday life.

Matter is the new ether.

Thought

Thought is a cross-section of order and time; self, of thought and body.

Thought is infinitely larger than the physical world. Practical thought is its finite portion.

Thought is as real as the physical world, but its edges are not straight, its sides are not square, and its corners do not come to a point. Thought is never seen or experienced by other observers. Thinking of thought, we find irony, enigma, anomaly, and paradox.

To think of thought is to poke the stick with the stick.

Yet we will poke at thought, with thought. We will sort things out even as they slide together or break apart. We will leap out of the dimensions — out of space, out of time, out of body, out of mind, leap out of the world that condenses above the cellular level. We will catch a glimpse of that world from the outside. We will appreciate how well dimensions work for multicellular consciousness and will venture, briefly, beyond them, looking back to see what dimensions are and why we think we are in them. We will see how practical thought is coordinated with the five dimensions of perceptual consciousness. We will see that thought is fundamentally temporal but that it evolves a spatial structure as it coordinates with doing in the physical world. Unlike other forms of thought (imagination, hallucination, spiritual experience, dreaming, etc.), practical thought has a dimensional structure, and for that reason coordinates with the physical world. Practical thought is not as clear and present as physical reality, but neither is it as ephemeral and illusive as non-dimensional thought. We will see that Dimension Six is still in the process of formation as it evolves in the present day. Science and technology are only now emerging as new forms of collective being.

The first five dimensions are entirely passive. We watch what is happening in space, time, and mass as if watching a movie: if the world were limited to five dimensions — to perception alone — we would be unable to do anything about what we see. But life is a movie in which we participate. Practical thought (dimensional thought that is closely related to observation) is what makes interaction with the world possible. If you think carefully of an order of things in spacetime that is physically possible — that is practical — you may be able, through the use of your body, to coordinate that thought with spacetime and do it — make it happen — in spacetime. You may be able to bring something you *think* into physical reality. Thought itself is not physical, but if it is practical thought, if it is coordinated with the physical, it becomes doable. Practical thought is similar to observation in that it must be something that could exist *physically*. Practical thought is different from observation in that it requires the body to interact with the physical world.

Before you do something, you think about it. It might be a split-second consideration, years in the planning, or a moment of caprice, but there is always an element of forethought in anything you do. It may be a barely noticeable ripple in the mind as you reach for the doorknob. You may fail to think properly, you may fail to think through to the end of a procedure, but you cannot fail to think before doing something. You conceive an order of things that does not exist and you bring that order into existence. You take a conceived object or pattern of objects from a realm of consciousness that is not physical, and you make it physical. The order you conceive before doing cannot be unbounded imagination. It is not anything you can think of — not riding to the moon on a bicycle or moving a sofa to the ceiling. It is what you can make happen in the world with your mind and body, and what you can see in spacetime after you make it happen. Practical thought has to be potentially perceivable, or it cannot be done. If it is improperly conceived, or incompletely thought through, it is not practical thought, and you will not be able to do it. It will not rotate into the physical dimensions. Practical thought is a realm of possibilities infinitely larger than, but dimensionally coordinated with, spacetime-mass.

Making a map is an example of dimensional coordination. Drawing designs on a piece of paper does not make a map. It is not a map even with roads and place names. But if you draw a picture with roads and place names that correspond with physical reality, you have created six-dimensional information — information that will rotate into spacetime as you drive down the road, and information that will become potential perception as you arrive where you are going.

Coordination, you will remember, makes interchange between dimensions possible. Because space is coordinated with time, we are able to interchange one time dimension with one space dimension — moving down the hall or along the highway — creating the sensation of moving "through" space. Similarly, the coordination of Dimension Five with spacetime makes it possible to interchange mass with time, or with one space dimension, to create acceleration and a "g force" throughout the body. You always feel the body, or part of the body, when you accelerate. Because practical thought is dimensional and coordinates with the other dimensions, it can, in conjunction with mass, be rotated into spacetime to create *doing*. To do something you have to think of a practical order and then put your body behind it. You have to push the furniture, or the pencil, or the keyboard to bring what you are thinking into space and time. *Doing* is a rotation of Dimensions Five and Six into Dimensions One through Four.

· · ·

Back in the room where we were before, the record player is playing the same old tune, the flowers in the vase are wilting, and you are getting tired of sitting near the table next to the same old lamp. You get up, walk across the room, pull the flowers from the vase, and toss them in the compost. Hearing Perry Como for the ninth time in a row, you think of Johnny Mathis. You find **Chances Are** *on the record shelf, take it to the record player, and put it on. Your mood lightens. Then you look over at the chair you've been sitting in next to the table and think of it in the empty corner of the room between the vase and record player. You don't see the chair there, but you would like to see it there, and it* **could** *be there. So, you walk to the back of the chair, and feeling its mass against your body, you push it across the floor. Once on the other side of the room, you turn the chair around and see it where you thought of it before. You have done something. You have actualized potential perception. You have rotated dimensional thought and mass into space–time. But rotation is not always possible. If you imagine the chair sitting on the wall, your thought would not have coordinated with the dimensions, and you would not actually see the chair where you thought of it.*

As you spin the chair around in its new location, your friend looks up from her chair and asks, politely, just what it is you are up to. You think for a minute and begin to move your tongue and diaphragm. "I just wanted to try it over here for a while."

. . .

Observational consciousness is envisioning an object in a dimensional context that someone is describing, where practical thought is thinking of what could be done in a dimensional context. Both are six-dimensional. *Creating* observational information is a form of doing: words and symbols are arranged in a practical order that creates a dimensional picture that anyone can experience observationally or perceive directly. Practical thought and observation are, therefore, the same realm of consciousness. The difference between them lies in the self, the combination of thought and body that makes doing possible. Observation does not require self—or doing—once the observational information is created. We will return to this topic later.

...

You are at the rifle range, this time with your friend. She has placed a paper target downrange with a glass bottle swinging on a string in front of it. You agree to try to shoot the bottle while wearing a blindfold. At times when the bottle is not swinging directly in front of the target, she has you fire a few practice shots. After the first shot you hear her yell, "Too far to the right." After the second one she says, "Too low." Then, after the third shot, she shouts, "Good! Now shoot there when I say so." As the bottle swings in front of the target you hear her say, "Shoot!" You pull the trigger again and the bottle shatters. You have located an object in observational space. When you take the blindfold off you perceive what remains of the bottle.

You could have seen the bottle all along had you taken the blindfold off. But you did not have to. You were able to "see" the target "through her eyes." With the help of language, potential perception was enough to locate the object.

When you and your friend are in the room, or when you are out front talking to your neighbor, with unicycles, dogs, and chocolate cakes all around you, physical objects seem to move in predictable patterns. When not subject to a force, each remains stationary or at a constant velocity. (Stationary is a constant velocity.) When subject to a constant force, each moves with constant acceleration. If you could measure (on the macroscopic level) the velocity and mass of each object, you could tell where and how each would move next. Objects seem trapped in a five-dimensional world of constant velocity and constant acceleration. But there are exceptions: a few objects in the *room and around the front yard change the rate or direction of their acceleration. These move in a noticeably different manner. Unlike chairs and tables, their motion is variable and unpredictable. They move around obstacles instead of colliding with them and move in an orderly manner towards objects with which they interact. Their motion could be described physically as* **orderly non-constant acceleration**. *These are observers. They are a type of object that you hear saying things about the world from their perspectives in space and time. You, your friend, your neighbor, and your neighbor's dog (and the squirrel) all move with orderly non-constant acceleration and talk — or bark or chatter — about the world they say they see. But is the dog an observer?*

You also notice that though these observers do not act with predictability, they act with **probability**. *You cannot calculate their future motions with certainty, but some future motions are more likely than others. Your friend is more likely to look your way than straight ahead when she waves at you; your neighbor is more likely to walk toward than away from you when he begins speaking from across the front lawn; the dog is more likely to run toward the squirrel than away; and the squirrel is more likely to run away from the dog.*

. . .

Observer behavior becomes more predictable in the aggregate. There is no way to know for sure what an individual observer will do, but in a society of several million observers, how many of them will drive on what highway on a given time of day is fairly predictable. Within a margin of error, one can say how many of them will stay home with a cold and how many will buy a pair of green socks at a department store. The more observers in the sample, the more predictable their aggregate behavior. You cannot say which observer will do what, but with enough statistics on aggregate behavior, you have a good idea as to how many people will be sick tomorrow and how many will buy green socks on Thursday.

The predictability of aggregate observer behavior and the probability of single observer behavior (which is its inverse) are understood against the background of potential perception: they might do this or might do that, but they could do anything physically possible. The future could be anything, but it is governed by waves of probability. As probability waves collapse into actuality, the future becomes the past. Potential perception collapses into actual perception. Probability applies at the rise of the collective from the individual level as it applies at the rise of the multicellular from the cellular level.

As the quantum is to macroscopic behavior, so the observer is to collective behavior.

A single consumer in a market economy is like a single particle in a quantized universe. You do not know what she will buy, but you know she is more likely to buy a sweatshirt on a cold day in Canada than a pair of sandals, and buy a beach towel on a hot day in southern France than a pair of snow boots. She could do anything, or nothing, so what she does against the parameters of what she could do is more of a wave than a point. But the high probability of some purchases under some conditions and low probability under others becomes the predictability of aggregate behavior in the market as a whole. We do not know if she is the one who will buy, but we can predict, within a range of certainty, how many sweatshirts will be sold in Canada on a cold day, and we know that number will be less on a hot day. We don't know who will be driving where or why, but on a given road we can predict more traffic at 5:00 pm on a Friday afternoon than at 3:00 am Tuesday morning.

Order always has a focus, or a point of view. Order from the point of view of a rat building a nest in your attic is not order from your point of view. An ant hill is always orderly from the point of view of the ants. A bomb detonating in your hometown is disorderly from your point of view, and orderly from the point of view of an enemy combatant. The focus of order is identifiable with an observer.

The "empty" potential of Dimension Six is *entropy*. Entropy is without order. It is a fire raging through a forest, an egg breaking on the floor, a rabbit crawling through a hole in the garden fence, a rock rolling down a hill—things happening without intention. It is the six-dimensional dispersal of five-dimensional energy. It is things falling apart: *disorderly* non-uniform acceleration. But entropy makes order in six dimensions possible.

It is what makes information possible. Entropy is the background against which order looks orderly—the more entropy in a system, the more disorder, and the more potential order. A blank computer screen has much more entropy than a telegraph key; there are millions of pixels at a single glance; each can be on or off and each can be a variety of colors. A telegraph key can signal only a dot or a dash and has much less capacity to relay information.

Order is life working against the arrow of time.

Order is the dimension that turns energy into work.

Evolution is possible due to the entropy of mutation. Perceptual information is possible due to the entropy of space, time, and mass, and observational information possible due to the entropy of Dimension Six. Information is the actual of order showing up against the potential of entropy. Stars and galaxies are the order we observe against the entropy of an expanding universe. The focus of order in the physical world as a whole is the source of religion.

There is order in the behavior of all observers, not just in those with information. There is order in the behavior of people, fish, bacteria, tigers, and earthworms, whether they communicate or not. As with the space dimensions, most of Dimension Six is devoid of actual information.

The order you experience in the shapes and sounds you see and hear in the behavior and speech of observers tells you more about the world than you could ever experience directly. The potential you experience in observational information is coordinated with the perceptual potentials the way they are coordinated with each other. That is why you potentially see or touch an object where and when an observer says she sees it. That is what makes entropy a dimension.

The Quantum Screen is grainy, but so finely grained that objects look smooth and continuous on the macroscopic level. Lines are straight; corners are sharp. Observational consciousness, though a dimension larger, is not as clear and detailed as perceptual. We "see through the eyes" of other people, but not as distinctly as through our own eyes because an observational picture comes in the form of words and numbers, not quanta. The observational picture is blurrier than the perceptual picture.

Science is a special form of observation—a more highly evolved form, as we will see. It is potential perception, like pre-scientific words and numbers, but clearer than words and numbers when experienced on the Pixel Screen. The Pixel Screen—the electronic retina of collective consciousness—is constructed by human beings to simulate the Quantum Screen. Pixels look like quanta. The Pixel Screen becomes a new realm of consciousness because it looks like the Quantum Screen.

As the Pixel Screen reaches higher levels of definition and availability, we will be less able to distinguish it from the Quantum Screen. We will look back and forth between screens the way we now "look" back and forth between seeing and hearing.

Thought is a finite cross-section in order and time.
In light it is an image in space,
in space-mass it is a word.

Self

If matter does not exist, there is only consciousness. But whose consciousness is it? Is consciousness the same as *self*?

If consciousness is not self, it belongs to no one.

Self is the axis of dimensional rotation.
Self is a finite cross-section of order, mass, and time.
Doing is self in space.

With Dimension Six we begin to do things. We decide things, move the chair, eat breakfast, go places we think of going, and do work. But before doing anything at all, we need to shape practical thought into a focus of order, or a sense of what does not exist but could exist; and we need to access the physical power to make it exist. We need to develop a self. Self is a shape in six dimensions: a physical body with length, width, depth, duration, mass, and order — a focus of order with a point of view — a mind with a body. A self is the axis by which dimensions rotate into one another, the axis of the five perceptual realms and practical thought. A self can be selfish — more individual than collective in its focus of order, or altruistic — giving of itself to a collective order. A self may be nearly all practical thought with little or no bodily potential, or all body and no thought. A self may have a clear plan but lack the capacity to rotate the right part of the body (arms, legs, tongue, fingertips) to bring thought into the physical world, or a self may be physically fit with no idea what to do or how to do it.

Doing is finite patterns manifest against an infinitude of six-dimensional entropy; it is self, imposing order on probability, creating objects and object systems where there were none. Future is the infinitude of Dimension Six — unlimited possibility with patterns of probability rotating into the physical world, with or without doing, with or without self.

Doing requires thought, body, and perspective in the physical world.

You are the axis of rotation between order-mass and spacetime.

Is a *self* the same thing as an observer? An observer, like a self, is a shape in six dimensions. Both observer and self move in orderly non-constant acceleration, at "right angles" to space-time-mass; and like selves, observers create order. Order in what they create distinguishes them from physical objects. Both observers and selves appear to "have" perceptual consciousness; the difference is that an observer is an idealized, selfless self that tells us what he is seeing, and where he is seeing it from, but does not bend what he sees to his own purpose. He doesn't do anything with what he knows other than tell you about it. He is a good scientist who reports accurately what he experiences and reports only what he perceives. A self, on the other hand, uses what he sees to do things. He mows the lawn, plays tennis, and supports his family. Your neighbor's dog is a self, but not much of an observer; he barks but doesn't say where anything is.

An observer communicates dimensional thought only, where a self communicates both dimensional and non-dimensional thought—imagination, hallucinations, ambition, opinion, hunger, desire, dreams, and spiritual experience. A self is a person. You can see her moving deliberately through space, avoiding objects, meeting people, moving her tongue and diaphragm in the air and her fingers on the keyboard, telling us what she sees and thinks, and focusing on what is important to her purpose. She does not pretend to be perfectly unbiased in her thinking; she does not pretend to see everything around her, as she focuses on what she wants to do in order to do it well. She will describe the red shoes she saw in the store an hour ago, but she is just as likely to tell you that she prefers the brown ones. What she tells you is not limited to potential perception. She's a little hard to get along with when she does not think to refill the ice cube tray, but she votes, pays her taxes, does volunteer work, helps a friend in need, and people generally like her. She is a citizen, a church member, and a volleyball player. She lives independently and takes care of herself. That is the focus of order that gets her to work every day, pays the bills, and keeps her alive. But her life is bigger than herself, despite her sense of independence: there are several larger collective selves — foci of order — of which she is a part. Her taxes keep the streets paved, her church feeds the homeless, and her volleyball team wins when everyone plays well together.

Order always has a point of view. Two runs in the fifth inning is good for your team and bad for the other team. Getting rid of cockroaches in your kitchen is good for you and bad for the cockroaches. A new car is an ego boost for you and a setback for a jealous neighbor. As gravity is the curvature of four-dimensional space-time into five, self is the curvature of five-dimensional energy into six. Self curves energy into work. It is a shape in six dimensions — an actual within entropy. Without the focal point of self there is no order, no Dimension Six, and no life.

A massive body bends space-time toward itself; self bends thought toward itself. To do things — to bend reality within spacetime-mass — self is restricted to what is physically possible, that is, to the realm of practical thought. To discover what is physically possible, the self constantly attempts to rotate the body, or parts of the body, into space-time. This can be as simple as pushing furniture around the living room or as complicated as moving fingers to type out a treatise on foreign policy. The self is constantly finding out what it can and cannot do. To be a shape in six dimensions, a self is less about what exists than about what could exist. It is not satisfied with what already is. Every living cell, organism, and organization is a self, and a self within a larger self.

Self is a curvature of thought, an axis of rotation, and a perspective in spacetime, but it is not consciousness as a whole. Consciousness transcends thought and the curvature of thought. This is important to everything asserted in this discussion. Many who consider the implications of modern physics know, at some level, that the physical world does not exist independently of consciousness, but they assume that without matter, self becomes the ultimate reality. This would be true if consciousness were the same as self. Ultimate reality would be solipsism, or the thought that, "Only I exist."

Spirit is consciousness unlimited by self.

Self is a finite shape within the infinitude of consciousness.

We share observational consciousness with other people, but not perceptual. They do not—you and I do not—"have" our own separate consciousness, perceptual or observational.

Does this mean there is only one consciousness?

Do only you exist?

It is true that you have no experience of other consciousnesses, but it is not true that you have no experience of other selves. Lots of other selves are out there, individual and collective, but no separate consciousnesses. Consciousness and self are not the same and each self does not own a consciousness. Consciousness is larger than your self and other selves, individual or collective, and larger than all selves together. There is a great deal a self is not and cannot do; but there is nothing that is not consciousness. Consciousness is not a thing within something else. Self is limited to what can be thought and done; consciousness is awareness of all the physical universe, all of life, and all of existence.

We spoke earlier of levels of consciousness: the cellular, the multicellular, and the collective. Among these, cellular sensation is limited to the chemical and tactile realms. As cells formed colonies and specialized into tissues and organ systems, some were specialized into sensory organs. Each cell retained a focus of order within its own separate chemo-tactile world. But as organisms developed complex systems of processing olfactory, auditory, and visual information, along with chemo-tactile information, cells became interdependent, and a focus of order developed around the cell community as a whole. Single cells retained, and continue to retain, a focus of order separate from the organism as a whole, but the organism has come to prevail over its constituent cells. With the perceptual realms, a self has evolved at a higher level of consciousness.

(The existence of order, and of what I have labeled *Dimension Six*, certainly existed during the evolution of the cell membrane, and earlier, which is to say, again, that there is no natural numeric order to the sensory realms and their corresponding dimensions.)

A parallel dynamic is now evolving within the human community. As the frontiers of science push ever outward, and more information comes through electronic media, the observational realm expands at the expense of the perceptual realms. We become increasingly interdependent as less of what we know comes through direct seeing and touching. Collective selves—religions, nationalities, professions, and football teams—become increasingly important. True, these are not new, but what is new is that through cameras, satellites, and cell phones we are beginning to see and hear at a much higher level of collectivity. A focus of order is shifting from the individual human to the collective, from the organic to the organizational level. A higher level of self is evolving in the realm of observational consciousness. You may still feel that "your" consciousness remains separate from that of others, but it is your *self* that is separate. The self retains a degree of separation from the collective focus of order no matter how highly integrated that focus may become.

First person self is actual experience in perception and observation.

Second person self is potential in perception and *actual* in observation. Second person is the self of collective consciousness and collective doing.

Third person self is potential experience in both perception and observation.

Does consciousness exist inside of other selves you see in the world around you? Most assume that it does but there is no way to know. There is a level of consciousness — the observational realm — over and above any of us separately, but is it in his or her self, in your self, or in my self? Is consciousness trapped within an organism or in any of its cells? Consciousness is best understood as not *in* others, or around them, or in cells, or anywhere else in space or time — not in the dimensions. Consciousness is indivisible; the self is what bends thinking to make consciousness seem divided and separate. This does not mean, of course, that other people are not conscious; it means that they are not separate. There is no separate perceptual consciousness inside of them. Through observational consciousness all people are united. We empathize with others through potential not actual perception.

· · ·

You are out back with your friend now, enjoying the afternoon air. Dark branches and light green leaves shimmer in the sunshine against a blue and white sky. A slight breeze rustles the trees, bearing a faint smell of mown grass. Words drift back and forth between you and your friend.

Her words coordinate with what you are seeing, smelling, and hearing. When she mentions to you what she is thinking, you listen, whether or not you agree, and usually agree once you understand what she is saying. The information in her words is more than information; it becomes a common vision. An equivalence exists between what she says and what you perceive. She bends your thinking to her point of view as you converse, but you feel a need to agree in an absolute sense only when the two of you decide to do something together. There seems to be something in her, in her eyes and how her face moves, that is in you, making you feel that your being is one part of both of you together. Her focus parallels your own — what happens to her happens to you, though you never perceive what she perceives. Equivalence of self — of your perceptual self with her observational self — creates a higher collective self.

The neighbor's back door slams and you hear the dog running to the front yard, barking angrily. When the neighbor appears across the fence you call out, "Hello."

"Nice out here," he says.

"Yes. Nowhere I'd rather be right now."

He's a friendly enough guy, but you don't know him well. He dresses oddly, leaves things lying around the yard, and has some questionable friends. He had a party that got a little out of hand a few weeks ago, the same day you had some friends over. The noise wasn't too bad — you were making a little noise yourselves — but you didn't like the tone of what they were saying on the other side of the fence and were glad when they left. Your friends were disturbed. The fence seemed to divide "their" collective self from "yours."

. . .

Caring for forests, oceans, and the atmosphere requires doing on the global scale: a universal scale that transcends nationality and ideology and includes all people. We need more observers to see further than we can see now, observers who see their own group in relation to others. We need to understand that the tools we depend on to preserve the separation of collective parts of the human family cut and maim the whole in their intended use. The paradigm of nationality, once crucial to human survival, is no longer adaptive. It will not live. We need a global self that can do.

Evolution proceeds in Dimension Six.

Evolution proceeds. The course of all life on Earth turns on the collective shapes humans create in Dimension Six.

Science

Consciousness is not an object of consciousness and cannot be known by science.

Consciousness has no preposition without paradox: beyond order, outside space, without mass, and before time.

The physical world does not cause or "enter" consciousness.

There is no observer-created reality.

The scientific method is a form of doing. Scientists constantly test practical thought by trying to rotate it into spacetime. That which rotates successfully is science, and that which does not, is not. The difference between scientific research and moving furniture around the room is in the self—in the focus of order. The moving of furniture is an individual doing for his own benefit, where science is doing on a level higher than individual self. The scientist is an observer—what he does is not for himself. By means of trial and error, he gathers information into an evolving universe of human consciousness unlimited by prejudice, preconception, and the perspective of his individuality. The scientific method creates a living picture of expanding observational consciousness.

Science is inclusive of all observers and not intrinsically curved, or focused, for the benefit of any particular individual or group of individuals. (A collective self can, however, curve scientific knowledge for its use.) Because scientific discoveries are observational information, they easily transform into practical thought and become doable by an appropriate "body." (Body is in quotes, because the agent of doing is more likely to be collective than individual.) What science learns is preconditioned for human manipulation, especially collective human manipulation—what we call *technology*.

The observational realm of words and numbers has existed far longer than science. For thousands of years, friends and neighbors have been telling each other about events and places across the woods, over the hill, in the next room, or on the other side of town. But prescientific observation does not present as detailed a picture as direct perception; information in words and numbers is coarse and inadequate when compared to the clarity of direct perceptual experience. Too often it is necessary to walk over the hill or into the next room to see things more clearly. To see things across town or around the world as clearly as direct perception would be a great convenience, a convenience that becomes possible in the present day through electronic technology. Television and computer screens provide working pictures of the larger world, pictures that are much closer to the clarity of direct perception, that simulate the structure of the Quantum Screen, and that are open to all observers.

I spend a lot of time outside. I walk, work in the garden, listen to birds, and look at trees. I like to know what the weather is likely to do. I look at the sky to see where clouds are forming and how they are moving. But now, with a cell phone, I am just as likely to look at the weather radar on the Pixel Screen. I used to think of the screen as a distraction from reality—a computer's idea of what is happening and rather than what is actually happening. But I have grown accustomed to looking at the phone as much as at the sky, and feel I am looking at the same thing either way. The two coordinate.

Weather radar is a moving picture of concentric green, yellow, and red shapes that begin an hour into the past and continue for three or four hours into the future. Green denotes light rain, yellow moderate rain, and red, heavy rain. For the hour in the past, where rain is known to have fallen, the shapes are irregular but well defined. This is the physical world — we know what has happened here. For the future, where rain is only likely to fall, the shapes are rounder and less defined. At the point where the shapes jump from the past through the present, there is a noticeable shift in definition. The meaning of the colors shifts from certainty of the past to uncertainty of the future and from intensity of rainfall to probability of rainfall. The yellow and red in the future still show the intensity — the probable intensity of rain — but the green shapes show the probability of any rain at all. They mean light rain or no rain. If a green patch passes over my yard, I may or may not get wet. I know for sure only when probability collapses into fact.

On the Pixel Screen, we are looking at the world collectively, from the sky.

The dimensional coordination of past and future is indicated by the similarity of shapes and colors as they jump across the time divide. As the weather forecast progresses ahead of time itself, shapes on the screen move from the certainty of the past to the possibility of the future. We assume the shapes we see are *the same things* in the past and future, though in the past they are objects and in the future probabilities. The two are coordinated, but only the past is physical; there are no physical objects in the future.

But does the weather have a focus of order? Is it a self? It appears entropic—purely random—but the fact that there is a level of probability indicates a degree of order, even if there is no apparent focus of order—no known intention.

Although scientists cannot measure order and cannot pinpoint its origin, some contend that order *emerges* from complexity, that complicated systems of interacting particles and force fields produce the order we see in weather (and in molecular structure, physical law, and life in general).

Non-scientists since the beginning of time have seen a focus of order in the weather, whether or not they observe it.

Model of Dimension Six: The Box and the Pixel Screen

The model for pre-scientific observation is the Box: rectilinear lines in space running through time. From the beginnings of language to the twentieth century, observational consciousness consisted almost entirely of words and numbers (with a few pictures scattered in).

The *Pixel Screen* is the evolving model of Dimension Six. As the retina facilitates the emergence of multicellular from cellular consciousness, the Pixel Screen facilitates the emergence of observational from perceptual consciousness. It does not change the object of experience; it changes the subject, from *me* to *us*. It is becoming an *electronic retina*, available to all observers. The Pixel Screen remains in a state of development; we experience it in scattered pieces: a computer on the desk, a television in the living room, a dishwasher panel in the kitchen, a cell phone in the palm of the hand. But already, we see less of the world directly.

· · ·

You are in the room once again, sitting on the chair where you moved it between the vase and the record player. The flowers are gone, but the lamp is still on the table.

You are looking around the room through the oval of photons. The table with the lamp is a perfectly flat surface with sharp edges and corners that come to a perfect point. What you see macroscopically is the **aggregate behavior** *of subatomic particles. Viewed up close their behavior is random and entropic; but from a macroscopic distance the wave function of each particle tends toward the flat, the sharp, and the pointed. The Photon Screen arranges the vase, the lamp, the table, and the record player into the space and time of perceptual consciousness.*

You also hear the din of traffic and the growl of your neighbor's lawn mower. Somebody is cooking cornbread in the kitchen. You do not see the lawn or the kitchen; you only hear and smell what is going on in these other places. The dimensions that structure the Photon Screen are extrapolated beyond the oval of what you see to what you hear, smell, and touch. The Quantum Screen extends infinitely in all directions beyond the Photon Screen, so you know where the traffic, the lawn mower, and the cornbread are without seeing them. And you know, through the coordination of dimensions that these are seeable where you hear and smell them. The information that becomes the surfaces and edges you see in the room is structured to coordinate with information in hearing, smelling, touching, and tasting. Someone talking about objects that you are perceiving in any realm—telling you that the cornbread is ready—is the coordination of Dimension Six with Dimension Five. Coordination at this level reinforces the illusion of objects waiting in space, independent of perception or observation.

· · ·

The transition of Dimension Five to Dimension Six (perception to observation) creates the sense of photons traveling through space. Your actual experience of photons—points of color (or gray) in a four-dimensional setting—is perceptual, but when you reconcile your point of view with that of another observer, the photon becomes a straight line through space, not a point. For scientific purposes the photon defines a line in space, even though space is defined by light. This is this circular construction that creates the paradoxes of modern physics.

. . .

Your friend is no longer in the room; she is on the phone telling you about the Grand Canyon. As she talks, she is hiking down the Bright Angel Trail toward Phantom Ranch. The sound you hear comes from your cell phone. This tells you where your phone is, whether or not you are looking at it. But the sounds of her words are ordered in such a way that you picture the scenery she is describing. This "picturing"—the Colorado River flowing between rock formations at the bottom of the canyon, the deep blue sky above the north rim, and the shadows lengthening on the trail—is the observational realm of consciousness. She says the trail is narrow and rocky, with passing groups of hikers and pack mules. She can see a reddish rock layer along the river at the bottom of the canyon and lighter sedimentary rock strata in the upper layers. She hears the descending twitter of a Canyon Wren.

You have never been to the canyon and are excited by what she is saying, but where do you put what she is saying? You can't just walk out the door to see what she is seeing, and you don't have time to go to the canyon yourself, so the picture you see is no more detailed than the words of which it consists. Your friend could try to describe the sharp edges of a rock *formation she is looking at, but the best she can do is to say it is "sharp" without showing the sharpness. You have a vague idea of the scene, and you sense her excitement in seeing it, but you envision only a rough dimensional arrangement of the hills and cliffs and plateaus that surround her. You know they are "real"—you know your friend is an honest observer—but there is no Quantum Screen on which to locate the things she describes. So, you put them in what I call the **Box**. The **Box** is a roughly six-dimensional universe in which you locate and store what you hear other people say about what they see. This is the prescientific observational realm of consciousness.*

But then you remember that it is the twenty-first century, and you ask your friend to switch on her video camera. You move your phone away from your ear and hold it in front of your eyes. Suddenly, there is the trail! You see the north rim, the river, and the blue sky on your phone's screen, right there in front of your nose. You see the red rock down at the bottom of the canyon with white streaks of igneous intrusion, and you hear the Canyon Wren through the speaker. Your friend points her phone toward a vertical rock edge that looks fairly sharp but is not entirely straight up and down.

The images you see and sounds you hear on your phone are the observational realm of consciousness in its scientific stage. It is far more detailed than the Box, and though smooth enough to make objects recognizable, still far grainier than the Quantum Screen. The screen on your cell phone is also limited to two dimensions of space (and one of time). You can see and hear (with the speaker) from the point of view of your friend, but you can't touch or taste or smell as she does. As currently available, the Pixel Screen on our phones and computers and televisions is a crude representation of the Quantum Screen, but we sense information on it the way we sense information on the retina. It's almost as good as direct perception and getting better every day.

. . .

The Box and the Pixel Screen are different models of the same realm of consciousness. They differ in detail. More importantly, they differ in their focus of order, and therefore in their sense of self. When you listen to what your friend is saying, you are hearing about the world from her point of view. No matter how honest and well-intentioned she may be, the information is filtered through her particular focus of order. If you do not know the speaker well, you will not know how honest and well-intentioned she is.

The Quantum Screen is a good model for perceptual consciousness because it demonstrates how thousands of separately activated points, or quanta, can create a composite picture on a higher level. Physical reality is a lot like the composite picture we see on a screen. But the analogy is really the other way around—the screens we have built to convey electronic information are built to be a lot like physical reality. Pixel Screens on our desks or in the palm of our hands are able to create actual experience because they mimic the composite structure of perceptual consciousness. The composite structure of the retina is re-created on the computer screen—that is how the pixel screen is able to become a realm of consciousness.

. . .

A fight has erupted on a street corner downtown. One man is hospitalized. Most witnesses admit that they did not see who started the fight; but one man, the brother of the man hospitalized, swears that it was started by the other fighter. Others say it could not have been. But a passerby with a cell phone happened to video the entire encounter, showing that the hospitalized man struck first. This is an undeniable Pixel-Screen recording of what happened—not a verbal construction—not a "he said, she said." Regardless of how fervently the victim's brother insists that the other man started the fight, we have a clear picture that everyone can see. The experience becomes collective. Within the context of a single realm of consciousness—Dimension Six—a new, collective level of consciousness in evolving through human technology.

. . .

Technology is biology.

View in Six Dimensions:

An object on the Quantum Screen is an object on the Pixel Screen; the difference is the subject.

The leap from multicellular to collective consciousness is a further evolution of self. As markets, elections, opinion polls, science, and electronic media shape information into a more concrete form, a new basis for collective self and collective action arises on the global level. The shape that the collective self assumes, united or divided, will determine the probability of human survival. The dynamic now — the *evolutionary imperative* — is a global self, born of scientific observation, firm in the spiritual equivalence of all human beings: a global self that can do what needs to be done.

To live, we must go beyond what we have ever done. We must see what is, what could be, and what is practical to do. We must create the world anew.

Spirit opens; self holds back, grasping, seeking, connecting, categorizing, focusing, curving, reducing, ordering — making practical use of experience.

Thought wanders, unfocused, striving for entropy, until harnessed by self.

It is spirit that makes me want to say what I see.
From the entropy of pure thought, self reduces spirit to word, and it is done.

Spirit moves self to do unto others as it would have them do. This is the transition to a higher level.

9

Summary

People are real.
Perception is real
Potential perception is real.
Imagination is real.

A physical object seems material when we see it and touch it. But we experience nothing beyond the seeing and touching. Matter is an assumption, a logical, common sense, assumption — we do not experience it directly.

Matter is a metaphysical assumption.

Physical science makes better sense without matter and multiple consciousnesses.

The dimensionality of mass obviates the independent existence of matter.

The dimensionality of observation obviates the multiplicity of consciousness.

The observational realm of consciousness is systematically constructed through the scientific method.

Collective self—that which does things above the individual level—is created not by science, but by the spiritual equivalence of perceptual and observational consciousness—by the equivalence of what you see and feel directly with what you hear other people say they see and feel. Seeing through the eyes of others on a global scale creates doing unto others and with others, on a global scale.

The growth of science makes global consciousness possible.

The growth of spirit makes global society possible.

Seeing stars and planets from a moving Earth is no different from seeing them from an immobile Earth at the center of the universe. The sky looks the same either way. An orbiting Earth becomes necessary only with our knowledge of physics.

Seeing dimensions as structures of consciousness is no different from seeing them as an external material universe. The world looks the same either way. Consciousness becomes the first principle of physical science only with our knowledge of quantum mechanics and relativity theory.

ABOUT SAMUEL AVERY

Samuel Avery holds degrees from Oberlin College and University of Kentucky. He has taught university courses in American History, European History, and American Government, and has been a builder, farmer, solar installer, and environmental activist. Avery has practiced meditation daily for over fifty years. He also has written a number of books (and audiobooks) on the relationship of physics to consciousness. Titles include: *The Quantum Screen*; *Dimensions Within*; *Buddha and the Quantum*; *Transcendence of the Western Mind*; and *The Dimensional Structure of Consciousness*.